Gradient Expectations

Gradient Expectations

Structure, Origins, and Synthesis of Predictive Neural Networks

Keith L. Downing

The MIT Press
Cambridge, Massachusetts
London, England

The MIT Press would like to thank the anonymous peer reviewers who provided comments on drafts of this book. The generous work of academic experts is essential for establishing the authority and quality of our publications. We acknowledge with gratitude the contributions of these otherwise uncredited readers.

This book was set in Times New Roman by Westchester Publishing Services. Printed and bound in the United States of America.

Library of Congress Cataloging-in-Publication Data

Names: Downing, Keith L., author.
Title: Gradient expectations : structure, origins, and synthesis of
  predictive neural networks / Keith L. Downing.
Description: [Cambridge, Massachusetts] : The MIT Press, [2023] | Includes
  bibliographical references and index.
Identifiers: LCCN 2022037237 (print) | LCCN 2022037238 (ebook) |
  ISBN 9780262545617 (paperback) | ISBN 9780262374682 (epub) |
  ISBN 9780262374675 (pdf)
Subjects: LCSH: Deep learning (Machine learning) | Neural networks
  (Computer science) | Conjugate gradient methods.
Classification: LCC Q325.73 .D88 2023 (print) | LCC Q325.73 (ebook) |
  DDC 006.3/2—dc23/eng20230302
LC record available at https://lccn.loc.gov/2022037237
LC ebook record available at https://lccn.loc.gov/2022037238

10 9 8 7 6 5 4 3 2 1

In memory of Julian Francis Miller

Contents

Preface

The main purpose of the brain, any brain, of any organism, is prediction—or so we are told by many prominent neuro- and cognitive scientists. I first encountered that claim in 2002, then again in 2005, and then again and again and again. By around 2007, I figured it was worth the time and effort to explore this fascinating hypothesis, so I started digging into the neuroscience literature—not an easy chore for a computer scientist. What I found were many areas of the brain whose structure and (purported) function made a lot of sense when viewed through *predictive glasses*, which I kept securely in place through many long hours of reading.

That personal journey led me to write two journal articles on systems neuroscience, with the brain's predictive machinery as a key focus. Unfortunately, each brain region appeared to be predicting in different ways, through complex interactions between a host of diverse neuron types and network motifs. As an artificial intelligence (AI) researcher, I found this particularly frustrating, since the primary AI interest in neuroscience has a very reductionist tint: we are a looking for *a few good neural principles* that mechanistically explain cognition and lend themselves to computer implementations (that ideally can plug-and-play in larger systems, such as self-driving cars, smart-home consoles, and so on). Neuroscience seemed to offer no such cheat sheet for intelligence, so I moved on to other endeavors.

In 2016, Andrew Clark (a popular philosopher among those of us who promote artificial life (ALife) approaches to AI) published *Surfing Uncertainty*, an enlightening account of the predictive mind. After digesting Clark's compact theory and the many examples that it convincingly explained, I knew that my return to the topic was inevitable. So in 2020, I began another lengthy investigation into prediction, but this time with the explicit goal of striking a healthy balance between neuroscience and connectionism, which, I quickly learned, had plenty to say about prediction, but only if you were willing to spend many hours absorbing the mathematics.

I invested those hours, and along the way it became clear that there was indeed a set of primitive computational mechanisms that both realized prediction in silico and, intuitively, presented perfect candidates for the building blocks of biology's expectation-producing strategies. Chief among these are *gradients*: relationships among factors that summarize the basic effect that a modification to one factor has on the other. From math class, we know these as *derivatives*: the change in Y divided by the change in X. Combining gradients with a few other simple operations, such as summing, averaging, and comparing, yields a small kit of versatile tools that support computational predictions, many of which can also be realized by simple combinations of neural units.

Not coincidentally, gradients also play a large role in the field of deep learning (DL), which has come to dominate AI over the past decade. They represent precisely the same concept in DL as in math class, but now their complexity puts to shame those triple-starred exercises in the final only-for-students-destined-for-a-Cal-Tech-PhD chapter of your calculus book. As a college professor, I spend several lecture hours explaining them to my (very bright) students. These gradients are the heart of DL; in fact, a common moniker for DL techniques is *gradient-based methods*. If you skip all the gradient calculations, you will never have more than a superficial understanding of DL. They are that pivotal.

The title of this book has multiple meanings. First of all, the blatant knock-off of a Dickens' classic seemed more appealing than *Predictions from Derivatives: Finally, Something Useful from Those Math Classes You Slept Through*. With *expectations* from *gradients*, the same idea should come to mind. So *gradients* have clearly earned a spot on the marquee, as have *expectations*, as a synonym for *predictions*.

This word combination has a second, equally important, connotation for this book: the expectations for DL's gradient-based methods are enormous. So much of the hype and hyperbole surrounding AI stems from the legitimate successes of DL, but most of us AI folks bristle at the mention of the many utopian and dystopian visions of a future dominated by our AI tools, many of which are envisioned as basic extrapolations of contemporary DL achievements. Although this book is certainly not an attack on DL—other authors have seized that gauntlet—it does call some of these expectations into question, particularly those concerning an *artificial general intelligence* (AGI) based on gradient methods.

Despite this skepticism to a continued domination of AI by DL, my foray into connectionism revealed many neural network designs from the late twentieth century that clearly accentuated the role of predictive machinery in cognition. Invented by some of the same people who drive the DL revolution in the 2020s, these older networks have not gone quietly into the night; their principles continually reemerge in nascent systems that try mightily to replace the complex, biologically unrealistic derivatives of DL with simpler, local (and thus biologically plausible) gradients. In so doing, these networks manifest the tight synergy between recognition (of sensory patterns) and prediction (of future patterns) that many view as fundamental to actual understanding. After all, we exhibit some of the deepest levels of comprehension by marrying our concept-identifying faculties with examplar-generating skills. Nobody has ever seen a mastadon on a putting green, but we can predict what such a scenario might entail; and by doing so, we reveal considerable deep knowledge of prehistoric mammals and golf courses.

The three components of this book's subtitle, *structure*, *origins*, and *synthesis*, refer to three primary subject goals of this work. First, the primitive predictive mechanisms can combine in many ways to produce predictive neural structures, as seen in both brains and artificial neural networks. The various primitives and resulting structures form the basis of the first three chapters of the book, with high-level conceptual explanations in chapter 1, more of the mathematical flesh and bones in chapter 2, and then the various neural implementations in chapter 3. Additional neural network structures appear throughout the book, including chapter 4, which delves into the older connectionist models mentioned above.

Chapter 5 focuses on *predictive coding*, a well-known principle dating all the way back to the 1950s, but still very prevalent in most discussions of predictive neural networks, several of which appear in this chapter.

The second goal, *origins*, takes center stage in chapter 6, which paints a picture of how predictive networks may have evolved, beginning with the simplest organisms and continuing on up through the mammalian brain and several of its subdivisions. This is far from a complete phylogenetic tree of predictive progress, but it helps set the stage for chapter 7, where the third goal, *synthesis* becomes the primary theme. My main interest and belief lies in evolutionary approaches to neural network design—a view that is anathema to many DL experts. Chapter 7 addresses the competition and cooperation between gradient-based and evolutionary routes to synthetic intelligence, before using some of the concepts from chapter 6 as the conceptual basis for an emergent predictive-network system that includes the three key adaptive mechanisms of this (and my earlier) book: evolution, development, and learning. The topic of synthesis arises throughout the book as discussions (both general and specific) concerning how various predictive techniques have been or could be implemented, but synthetic issues reach a head in chapter 7. Finally, chapter 8 summarizes the earlier chapters, derives several generalizations from them, sketches a few predictions of its own for the future of AI, and then boldly refuses to pick sides.

This book's target audience is anyone with a deep interest in intelligence and how neural structures might achieve it. This, quite naturally, pertains to college students who study psychology, neuroscience, or AI; but the main concepts should be accessible to anyone with an interest in cognition and the patience to follow my virtual dissections of neural networks, both simple and sophisticated, biological and artificial.

Although mathematics appears throughout the book, only chapter 4 burrows into it very deeply, as I try to do more than just hand-wave at the ties among crucial concepts such as prediction, recognition, information, surprise, and free energy. However, for those who trust a hand wave, the many gray boxes filled with derivations can easily be skipped with no repercussions in later chapters.

Prediction is a popular topic, of which many books have been written. My original thought was, essentially, *Who needs another one?* But as I read through many of them, I found myself wanting more concrete mechanisms, more of the dirty little details that make things resonate in my own mind. So although this book is hardly the definitive manual of a prediction machine, I hope it gives you a feel for how the high level expectations of mice, macaws, and men ground out in patterns of neural activity . . . and how scrutinizing these vertical slices, from the coarse spatiotemporal scales of ethology down to the microns and milliseconds of neuroscience, might help us improve the intelligence of our machines.

Acknowledgments

Back in the early 1980s at the beginning of our afternoon runs for the Bucknell cross-country team, we would occasionally hear the chant, "FUBAR, FUBAR, FUBAR," from several teammates. At the risk of losing grade-school readership, I will only repeat the final three words of that acronym: beyond all reason. The call for FUBAR was a desire to let our whims lead us in any and all directions for those 70–90 minutes of running, whether across a superstore parking lot, down the rows of a cornfield, or through a carwash. Very little was out of bounds as no-trespassing signs became mere recommendations for our own safety.

My own research in artificial life (ALife) and artificial intelligence (AI) has always had a FUBAR flair. I can change directions whenever some shiny object pops up that exhibits an interesting form of life or intelligence that seems amenable to computer simulation. When one of those attractive baubles turned into a lump of coal back in 2020, a few months into the pandemic, I did a little soul-searching and decided on an abrupt shift and returned to the concept of the predictive brain to see what more I could find in the neuroscientific, and particularly the connectionist, literature. Uncertainty was the only certainty in that pursuit, so I contacted Elizabeth Swayze at MIT Press to try to get some indication of whether a book in that area was a worthwhile goal. Her encouragement was a much-needed and appreciated constant for the ensuing 15 months of research and writing. Matt Valades entered the process a little later and was equally helpful. Now, as my part of the project draws to a close, Theresa Carcaldi stands by to clean up my grammatical messes with her eagle-eyed copyediting. I am very grateful to all three of these patient and supporting individuals.

I once gave a lecture on AI and ALife to a group of engineering professors. One of their questions, apparently meant as an insult, was, "Is this stuff science . . . or art?" Coming from the liberal arts side of mathematics and computing, I found the answer quite obvious: it's both. A good many researchers in the sciences of the artificial have a creative passion (often bordering on artistic) that drives the field. When that combustible, visionary energy and the untethered explorations that it inspires gives way to so-called epsilon research and citation counting, it's time to look elsewhere for fulfilment. For me, book writing provides the perfect opportunity to explore fascinating topics and mix a bit of art with the science without fear of repercussions from just-the-facts, competitive-results-only journal reviewers.

Although Wolfgang Banzhaf encouraged me to take a shot at long-form writing over 20 years ago, it took more than a decade for me to summon the gumption to follow through.

I am very grateful to Wolfgang for his continuing support of my research and writing. My former and eternal Bucknell teammate Eric Allgaier originally convinced me that my words were worth reading, and to this day he serves as my unofficial publicist as he flashes copies of my texts during group Zoom calls and jokingly recommends my materials as excellent bedtime reading and Christmas gifts to our friends.

In March 1993, Jim Valvano, one of college basketball's most famous coaches, gave an unbelievably sad yet gripingly inspiring speech just a few months before dying of cancer. The slogan from that evening and for the V Foundation for Cancer Research is simple and poignant: "Don't give up; don't *ever* give up." Although many of life's adversities are trivial in comparison, those words have motivated me in so many instances, including the writing of this book. There were many points at which I saw no easy road forward, but then a second phrase, by one of college football's most famous coaches, Woody Hayes, would start to ring in my ears: "Anything that's easy ain't worth a damn." Finally, my own college coach, Art Gulden, supplied daily injections of motivation that are not encapsulated in a single most-memorable event or quotation, but his effect upon the lives of many was very profound and perfectly in tune with Valvano's message of unwavering perseverance.

In July 2001, I presented two completely different and fully FUBAR papers at the Genetic and Evolutionary Computation Conference in San Francisco. After the second talk, a soft-spoken yet highly energetic man, Julian Miller, greeted me in the hallway and expressed his interest in my work. That encounter started two decades of informal collaboration and spawned many research visits of Julian traveling to Norway, and me and my Norwegian colleagues traveling to England. Julian became very prominent in the fields of ALife and evolutionary computation, so I have always considered it a privilege to share a certain scientific wavelength with him, even with the diverging trajectories of our careers. My continued optimism for the ALife-based methods discussed in chapter 7 of this book, regardless of the onerous challenges, directly stems from watching Julian doggedly attack them for 20 years with no hint of regret nor fear of failure despite the long odds. Julian lost his battle with cancer in 2022. This book is dedicated to his memory.

You will have a hard time finding the FUBAR philosophy promoted by any of the normal research funding agencies, despite the fact that a large percentage of funded research is exploratory and leads to no noteworthy breakthroughs. Hence, for the nearly 40 years of my career in AI and ALife, I have had no significant association with any of these sources. I advise my share of PhD and master's students, but I do my best to shield them from any negative consequences of my wild ideas. The vast majority work on relatively safe and straightforward projects that will lead to a degree. These days, my FUBAR runs tend to be solo. I owe a great deal of debt to the Norwegian university system, which still allows professors enough time and basic resources to pursue creative passions (albeit with constant encouragement to join large international projects), and gives reasonable teaching loads that make it possible to devote oneself to producing a few high-quality courses instead of a bevy of mediocre offerings.

At the local, departmental level, my attempts at book writing have received stout backing as a legitimate form of research and scholarship from a series of talented department heads: Guttorm Sindre, Jon Atle Gulla, Letizia Jaccheri, John Krogstie, and Heri Ramampiaro. I would have never realized my primary professional goals without their generous support and understanding.

I would have never even logged on without the frequent assistance of Erik Houmb, our systems ace who has guided me through the jungle of online tools for the latter half of my academic career. Unlike nearly all my family members, Erik has resisted the (surely gnawing) temptation to sarcastically inquire, "*You* have a PhD in computer science?"

My friends in the neuroscience department at Oberlin College also deserve a round of applause for their assistance. Not only do they give me a place to hang my thick, Norwegian wool hat and connect my laptop during sabbaticals, but they trust in my ability to teach and advise their (very bright) students despite my lack of a formal neuroscience education. I hope to visit you all again soon, Kristi Gibson, Gigi Knight, Mike Loose, Tracie Paine, Pat Simen, and Jan Thornton.

When working in Oberlin, I receive bottomless generosity from my Ohio family in many departments, from transportation and lodging to dining and entertainment. Though I work very hard during those visits, each day still feels like a vacation thanks to them. My brother Steve Downing, my sister-in-law Meg Downing, and my cousins Kathy and Denny Mishler are all vital links back to my primary identity as an American citizen ... and Cleveland sports fan.

My wife, Målfrid, has repurposed the FUBAR run to the FUBAR hike, with the latter taxing my muscles, joints, and psyche much more than any college cross-country workout. By yanking me from my comfort zones of groomed trails, bike lanes, and swimming pools, she has opened my eyes to a beautiful world off the beaten path, even when the unbeaten path is thinner than its corresponding line on the map, and I stumble along, far behind the group. I wish I could cite Jimmy Buffett and say, "With you I'd walk anywhere," but she knows that to be true solely in the metaphorical sense.

Our children, Neva, Asta, and Jasem, help keep us young. They inspire us with their diligence and steady stream of hard-earned successes, whether in music, sports, or academics. But no prizes, medals, or scholarships make us prouder than simply observing the kindness that they show to others.

Finally, I owe an immeasurable intellectual debt to my PhD advisor at the University of Oregon, Sarah Douglas, and to two professors in the Department of Mathematics at Bucknell University, Eugene Luks and Michael Ward. Mike awakened me (literally, in 8 am classes) to the intricate patterns and rigorous proofs that make mathematics such a gratifying pursuit, while Gene cultivated my appreciation for computation as an abstract and omnipresent process, whether he intended to or not. Sarah grounded artificial intelligence in cognition, thus magnifying the appeal of both subjects. Their aggregated influence stoked my endless curiosity as to how it all fits together. This book is one of the more personally satisfying by-products of that quest.

1 Introduction

It's tough to make predictions, especially about the future.
—Yogi Berra (famous American baseball player and coach)

Aside from the classic prerequisites to evolutionary success—survival and fecundity—the ability to predict clearly tips the Darwinian scales like few other cognitive faculties. Having just an inkling of what lies around the bend, behind the bush, or over the horizon can spell the difference between feast and famine, pleasure and pain, life and death—or, in Yogi Berra's world, strikeout and round-tripper.[1] People who know what lies ahead can amass fame, fortune, and a long line of eager followers and envious competitors.

But despite those common shortcomings that render the bulk of us losers in Las Vegas casinos, suckers for poor investments, and benchwarmers in baseball, most of us possess the predictive apparatus that aided man's ascent to the top of the food chain; and that, if nothing more spectacular or profitable, does help us snake our way to a speedy checkout at Food Lion. We may not know whether pork bellies will trade higher or lower tomorrow, but we can easily surmise that the three teenagers, each with a single bag of pork rinds at register 3, will file out long before the guy with a full cart and a screaming baby in the express line, or the partygoer in line 7 with only a few items but no visible means of differentiating a debit card from a driver's license. When paying attention, we make some quick, predictive calculations, jump to aisle 3, and never look back. We hardly even recognize that we've done any complex thinking. After all, it's not rocket science . . . unfortunately.

In 1969, when Neil Armstrong took his giant leap for mankind, would anyone have ventured that, a half century later, the final frontier would be between our own ears? Intelligence exemplifies Churchill's "riddle, wrapped in a mystery, inside an enigma," and somewhere buried deep in that tangled mess lies prediction, not as a disjoint, free-floating feature but as the featured attraction.

In the past few decades, many prominent cognitive scientists (Llinas 2001; Hawkins 2004; Clark 2016; Buzsaki 2006) have begun to hail prediction as the hallmark of intelligence. For example, Buzsaki sets a predictive tone with the first sentence of *Rhythms of the Brain* (Buzsaki 2006, vii): "The short punch line of this book is that brains are foretelling devices, and their predictive powers emerge from the various rhythms they perpetually generate."

Equally convinced is Llinas, who writes, in *i of the Vortex* (Llinas 2001, 21), "The capacity to predict the outcome of future events—critical to successful movement—is likely, the ultimate and most common of all global brain functions."

Complementing those two prominent neuroscientists, the cognitive scientist and philosopher Andy Clark prefaces his popular *Surfing Uncertainty* (Clark 2016, xiv) with his usual elegance:

The mystery is, and remains, how mere matter manages to give rise to thinking, imagining, dreaming, and the whole smorgasbord of mentality, emotion and intelligent action. . . . But there is an emerging clue. . . . The clue can be summed up in a single word: prediction. To deal rapidly and fluently with an uncertain and noisy world, brains like ours have become masters of prediction—surfing the waves of noisy and ambiguous sensory stimulation by, in effect, trying to stay just ahead of them.

No less emphatic is tech entrepreneur and cognitive scientist Jeff Hawkins, who writes, in *On Intelligence* (Hawkins 2004, 89), "The cortex is an organ of prediction. If we want to understand what intelligence is, what creativity is, how your brain works, and how to build intelligent machines, we must understand the nature of these predictions and how the cortex makes them. Even behavior is best understood as a by-product of prediction."

Although the advantages of explicit predictive skills in everyday life seem obvious, the proposals by the above scientists are much more radical, as they argue for prediction's centrality in the workings of the brain. This book investigates that claim by examining a host of neural networks, both natural and artificial, with a special focus on the internal flow of signals that seems to embody expectations.

1.1   Data from Predictions

From a practical machine learning (ML) perspective, the ability to predict provides an invaluable service to *data-hungry*, supervised-learning algorithms, such as most conventional neural networks. The well-known fuel of these algorithms is data—as in the popular phrase, *data is the new oil*. For a supervised-learning system, data constitutes pairs of input patterns and their corresponding target output patterns (often called *labels*). For example, the data pairs for a facial recognition system consist of pixel images labeled with names, social-security codes, or other unique identifiers. In today's online world, there is no shortage of images, text, and other unstructured data, but labeling often requires time-consuming human analysis.

This same human bottleneck applies to supervised sequence-completion tasks, in which a system receives several words, phonemes, images, or other unstructured elements of a sequence and must *predict* the next item. In these cases, the next item constitutes the target, and when these targets are full images, acoustic patterns, and the like, the labeling chore for humans becomes all the more arduous.

The beauty of predictive algorithms is that if they operate somewhat autonomously in a *situated setting* (i.e., they have direct access to a physical or virtual environment in which they can move about and sense their surroundings), then they generate their own targets by merely *waiting one timestep*. Any prediction, made at time T, of the environmental state at time T+1, will receive its target when the agent observes the world at time T+1. Thus, the data item is the state at time T (plus possibly other states prior to T) paired with the (target) state at time T+1.

By moving around, observing, and predicting, the agent generates its own data set, which it can then use in a supervised-learning fashion to improve its own predictive abilities. And improving one's competence at predicting future states of the world often equates with building an accurate model of both the world and the effects of one's actions on the world, which are, in turn, two core aspects of intelligence.

1.2 Movement and Prediction

The most convincing argument for prediction as the primary prerequisite to intelligence revolves around movement and its pivotal contribution to survival. Consider the agent of figure 1.1 and its ability to respond to environmental change. In this diagram, the time required to sense and then act generally undercuts the duration of an environmental state, i.e., the environment's timescale (τ_e) exceeds that of the agent (τ_a). The sequential processes of reading world state X and responding to it fall within X's time window, thus making the response appropriate.

Conversely, when the agent operates at a lower frequency than the environment (as in figure 1.2), the delay between sensing and acting causes obvious problems: the agent senses state X and responds to it, but only after the state has changed to Y, yielding the response inappropriate.

Faced with an environment that changes faster than the latencies of sensing and acting allow it to respond ($\tau_a > \tau_e$), the (perpetually confused) agent's prospects for a long and happy life seem grim.[2]

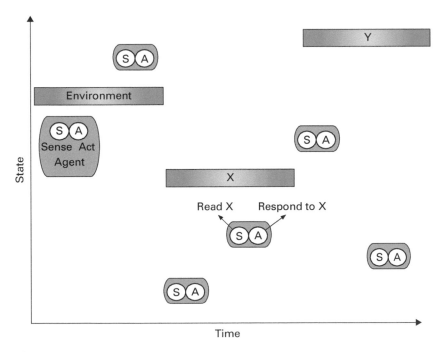

Figure 1.1
Comparing the natural timescale of a sensing-and-acting agent (τ_a) to that of its environment (τ_e) when $\tau_a < \tau_e$. Horizontal bars denote relatively stable environmental states, while the agent's vertical position denotes its own state. Larger time constants entail longer stable states.

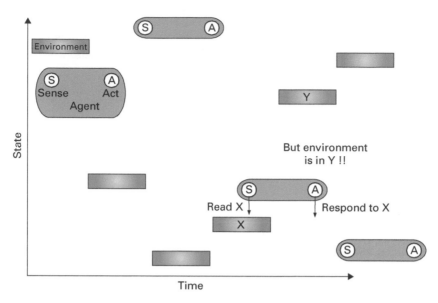

Figure 1.2
Comparing the natural timescale of a sensing-and-acting agent (τ_a) to that of its environment (τ_e) when $\tau_a > \tau_e$.

Prediction provides the perfect antidote to this temporal mismatch and indecision. As shown in figure 1.3, a predictive agent can sense state X, use its model of the world to predict that X leads to Y*, and then respond to Y*. As the figure indicates, Y* only approximates Y, but anything close to Y supports better preparation than the assumption that X will persist. It is probably better for a gazelle to run away from a rustling bush with lingering doubt as to the exact angle from which the tiger will pounce than to ignore the possibility of an abrupt and violent change of state.

Interestingly enough, the gazelle's innate speed compounds its predictive challenges. The faster an agent moves, the more quickly its environment changes. This motion-induced environmental timescale (τ_m) essentially supersedes τ_e in estimating the agent's evolutionary fitness. The faster it moves, the smaller τ_m becomes, and thus the greater the need for low sense-and-act latency and/or accurate prediction. Basic properties of biochemical signaling and neural circuitry place strict lower bounds on latency (τ_a) such that the only feasible solution for almost all mobile organisms involves predictive mechanisms. As Llinas (2001) argues, these predictive abilities are the brain's main function, and to such a convincing degree that primitive sessile organisms need no brain at all. He uses the (now popular) example of the sea squirt, which begins life as a free-swimming larva before permanently attaching itself to a fixed location and digesting its own brain, which would apparently serve as nothing more than an energy sink during its adult stage.

Llinas extends his argument to frame cognition as internalized motion control. Early in embryonic development (as well as in the mature stages of very primitive organisms), the activity of muscles is controlled locally, in a very emergent (but limited) manner: active muscles stimulate neighboring muscles, often in a rhythmic manner, which serves many useful purposes in both movement and digestion. As motor neurons arise and their axons migrate and connect to muscles, that control moves upward in the neural hierarchy. Two key

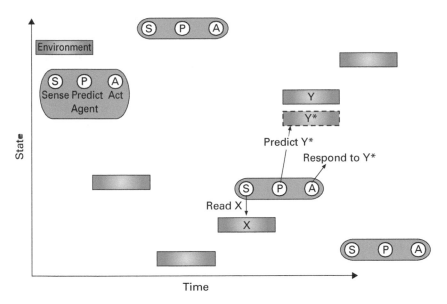

Figure 1.3
Adding prediction to the agent's cognitive repertoire and thus combating problems associated with timescale discrepancies: $\tau_a > \tau_e$.

advantages of this *encephalization* are (1) coordination of more intricate muscle activity by neurons that are both intraconnected in complex patterns (more than are the muscles themselves) and linked to multiple muscles, and (2) creation of avenues for integrating diverse sensory inputs from throughout the body into the decision making of the premotor neurons and higher-level circuitry. The price paid for this complexification is latency, which rises by tens to hundreds of milliseconds per neural layer. Hence, the need for sophisticated sensing becomes acute, as a slow decision process needs a more accurate *picture of the world* and, even more important, a model of the agent-world coupling, in order to make reasonably accurate predictions of future states and to coordinate a platoon of muscles to operate in an elaborate environment. The complexities of sensing, acting, and predicting must increase in lockstep. Thus, to the extent that evolution bootstrapped motion control up the neural hierarchy, prediction needed to join the ascent.

1.3   Adaptation and Emergence

Adaptation is a system's ability to change in response to (or anticipation of) changes in its environment. The above gazelle example fits this definition, as does the ability of herbivores to evolve faster reaction times in their arms race against carnivores on the African savanna. Adaptation spans many timescales.

In this book, the term will be restricted to internal *structural* change of some significant duration (typically in a neural network). So the change in a gazelle's speed and direction at any given moment will not fall under this more-restrictive definition, but a change in neural synapses (their number and/or strength) to modify the animal's conception of *dangerous bush-stirring sounds* would qualify as a relatively short-term adaptation.

Carving the timescales into three general pieces (short, medium, long) results in three standard classes of adaptation: learning, development, and evolution, with some overlap among them, especially the former two. However, given our focus on the brain and other (artificial) neural networks, a reasonable working distinction is possible: learning involves changes to existing interneural connection strengths, while development involves the formation of entire network topologies, including the formation of connections. Granted, in real brains, new synapses grow and die throughout life (as part of learning, disuse, and so on), but the net result is the change in one neuron's effect on another. Those changes of influence versus changes in topology will be the main distinction between learning and development in these chapters.

Evolutionary change is easier to distinguish, as it requires inheritable genomic modifications (for the most part): structural change is to the genetic material that gets passed on to the next generation. An evolutionary change is thereby recognized as a structural difference between an organism and its descendants.

Although changes to the immediate firing levels of individual neurons typically reflect responses to those of other neurons or to environmental stimuli, that timescale of adaptation will typically not fall under our working definition, since the normal changes inherent in neuronal firing will not be considered structural.

The processes directing adaptive change have great significance in this book. Those involving a central control algorithm that analyzes many or all components before adjusting each such unit are of only peripheral interest. My focus is on emergent systems in which local activity leads to global patterns, which may eventually exert some influence on the local behavior. But those global influences are not hardwired into the system; they must arise from the local dynamics.

My earlier book, *Intelligence Emerging* (Downing 2015), delves deeply into emergent mechanisms underlying learning, development, and evolution with respect to natural and artificial neural networks. In that work, I posit that each emergent adaptive process in nature appears to be driven by relatively random trial-and-error search for appropriate synaptic strengths, stable and efficient network topologies, and high-fitness genomes (which are genomes that, among other things, encode good recipes for development and learning). Crucially, the trial-and-error processes at the slower timescales produce landscapes for the search performed by the faster adaptive processes. Even the finely tuned (by learning) network of synapses sculpts basins of attraction that strongly bias the dynamics of reasoning.

Although *Intelligence Emerging* includes some limited material on predictive mechanisms in the brain, the focus is on the trial-and-error processes by which neural motifs may learn to encode predictions. Explanations of how prediction *grounds out* in stochastic search removes any serious contradictions between that book and this one, but the tension between those two underlying mechanisms of intelligence deserves careful scrutiny.

When I have expectations about possible futures and leverage them to choose actions, clearly I am not following a pure trial-and-error process. The (albeit uncertain) lookahead governs actions that are far from random. Information about how tweaking parameter A will affect component B gives any agent an advantage over those who will randomly change any parameter in the quest for improvement.

Clearly, the emergence of intelligence, from bacteria to humans, cannot be explained by purely random activity, but at each temporal scale, processes governed by seemingly arbitrary *choices* play a strong role. At the evolutionary timescale, the unpredictable (though not purely random) mutations and recombinations of genetic material produce new genotypes, which are then subjected to the decidedly nonarbitrary forces of natural selection. Over the generations, the genetic material figuratively *learns* the effective adaptations for a given environment. In infants, random movements dominate as the brain gradually sorts out what works and does not. In general, all species in all phases of life strike a balance between *exploration* and *exploitation*: relatively random activities versus those known to be productive, where the former can be viewed as *reconnaissance* expeditions, with the accrued information used to guide later exploitation.

Intelligence Emerging puts extra emphasis on exploration, due largely to my fascination with the nondeterministic (yet doggedly persistent) nature of so many biological processes. This book digs deeper into exploitation, via prediction. However, emergence remains a dominant force in the chapters that follow.

1.3.1 Gradients and Emergence in Neural Networks

The critical divide between the neural networks covered herein and those that have taken the AI world by storm since the early 2010s is the presence of a global control mechanism, and thus lack of convincing emergence, in those popular neural architectures. The vast majority of successful deep learning (DL) networks employ effective global controls powered by information with extensive spatial scope: causal knowledge of how tweaks to component A will affect the behaviors of other components, some of which may be quite distant from A in structural space, aka *gradients*. This spatial lookahead arises not by trial-and-error experience but by formal mathematical derivations across long causal chains. These provide computational templates into which the data of individual experiences nicely fit, yielding precise numerical gradients and thus well-founded lookaheads for intelligent action selection. For example, the question of how to most judiciously change one of the millions of weights (w) in a deep network so as to reduce the total error (E) on the output end of that network is answered in DL systems using the gradient $\frac{\Delta E}{\Delta w}$, which represents the expected change in E due to a unit change in w. A DL system computes one such gradient for each of the millions of weights. Thus, it amasses detailed information that helps predict how the change in any particular weight will contribute to the ultimate goal of reducing E.

Gradients play a central role in this book as well, but they have a much more local property: they link changes in one variable (e.g., the strength of a synapse) to *nearby* changes, such as to the error or *surprise* recorded by a neuron immediately downstream of that synapse. Other gradients capture differences in the firing rates of neurons across short expanses of space or time, again, local relationships and computations. Another variation of gradient records the changes in the predictions themselves, with larger differences manifesting *surprise* and stimulating learning. In each case, these local gradients and their usage exemplify emergence and often reflect current trends in neuroscience. Conversely, the *long-distance gradients* of contemporary deep learning have a much more tenuous relationship to biology, despite the fundamental biological inspiration of neural networks in general.

As a simple analogy, consider person X running for president of Land-O-Plenty. The ideal scenario for X is to know, for each citizen, c_i, how changes in c_i (with respect to acquired information, income, services, and the like) will ultimately affect the total number of presidential votes that X receives (V_x). That is, X would like to know $\frac{\Delta V_x}{\Delta c_i}$ for each individual i. This is a huge request, but one that has, for better or worse, become reasonable in the social-media age. X can then target each voter with surgical precision, feeding them just the information (real or fake) that X can predict (with reasonable certainty) will nudge c_i toward voting for X (and broadcasting praise of X to her contacts). In short, X computes $\frac{\Delta V_x}{\Delta c_i}$ for all 350 million citizens, c_i, and then uses that as the basis for 350 million individual predictions as to how c_i will probably vote (and influence other voters) given a particular change Δc_i that is tailored specifically for c_i.

Each of these gradients will be very complex, involving detailed reasoning about c_i's age, education, employment, lifestyle, voting history, and so on such that the logical connection between the tailored information and the desired support is long and winding. For example, telling c_i that *my opponent, Y, wants to guarantee funding for a nice new stretch of highway between Barleyburg and Soy City* will resonate poorly with c_i, since she (a) runs a popular general store on a scenic back road that is currently the only connection between Barleyburg and Soy City, (b) has no formal education beyond junior high school, and (c) is approaching retirement age. For the well-funded and tech-savvy political campaign, this type of information may lie within its reach. For the backpropagation algorithm that drives deep learning, it is a prerequisite, and one easily obtained with enough computing power.

In earlier days, before people eagerly divulged troves of personal information in public fora, X would have had to get by with knowledge of other, more general, gradients, such as $\frac{\Delta c_j}{\Delta c_i}$ for *all* friends, that is, the same relationship holds for any pair of friends, i and j. For example, as a heuristic (i.e., rule of thumb), X might assume that if c_i tells friend c_j something that c_i has just learned and now believes, then c_j will believe it too. In short, X understands how local changes in the beliefs of people come about: how information and belief spread *through the grapevine*. X can then plant the seeds for spreading belief by taking out attack advertisements in local newspapers against candidate Y, in the hopes that $\frac{\Delta c_j}{\Delta c_i}$ will kick in numerous times, producing widespread belief in Y's incompetence and evil intentions. Alternatively, X could promise (if elected) to pump government aid into the local economy, hoping that word of this favorable action would quickly reach all citizens of Barleyburg.

Basically, X must rely on global or regional messaging (or money) and a general intuition about local causal relationships (gradients) in the hopes of producing a victorious global outcome: $V_x > V_y$. But X has no *direct, long-distance line* that links c_i to V_x: a single individual to a global outcome. This is politics the old-fashioned way, and learning the biological way.

Brain areas can broadcast global or regional signals (i.e., neuromodulators), but the only pinpoint messaging occurs between a neuron and one of the (possibly ten thousand) others to which it connects. The history of that local signaling plus any nascent or lingering neuromodulator then drive synaptic change, the cornerstone of learning. There exists no brainwide ledger that predicts how any synaptic change will affect overall behavior. Emergence, with all of its uncertainties, is the only known biological route to cognition and survival.

In summary, today's most powerful deep-learning systems rely on long-distance gradients and thus exhibit very little emergence, despite the fact that a researcher can rarely predict what such a network will learn from a bevy of examples. These nets are still surprising and normally frustratingly difficult to explain, but that complexity does not arise from purely local interactions. However, biological intelligence, with its inherent adaptability, does rely on local interactions and emergence across multiple timescales. And as shown in the chapters that follow, many of these local mechanisms embody prediction when one carefully examines the nature of expectations in a neural system.

1.4 Overflowing Expectations

This book will not help you beat the stock market, prepare for tomorrow's weather, or move more efficiently through mega-store checkout lines. Sorry. The final, overt predictions produced by humans and machines are really only of peripheral relevance to the main story: expectations seems to be omnipresent in the brain, and our AI technologies could benefit by incorporating similar, widespread, local predictive mechanisms into neural networks.

At one time in our distant evolutionary history, the overt behavioral outputs were the main predictive achievement, but as nervous systems arose and evolved, the predictive capabilities ascended and proliferated throughout the neural circuitry. Today, many regions of the nervous systems of numerous animal species are amenable to useful predictive interpretations, often involving the collision of a *reality stream* and a *prediction stream*, the difference of which yields a *prediction error*, which constitutes a modified reality stream that continues upward in the neural hierarchy, while also contributing to a prediction of its own about activity at a lower level.

Under this interpretation, the brain is a flood of expectations and surprises (violated predictions) gushing down and up (respectively) the neural hierarchy. This view of neural processing has been around for more than seventy years and gained significant popularity over the past few decades. The intuitive advantage of such an arrangement seems pretty obvious: if information is *as expected* by some receiving region, then why should the sending region expend the energy to transmit it? When nothing unexpected happens in downtown Soy City, journalists resort to human-interest and nostalgia pieces to fill the local paper. The brain has no problem with slow news periods; it can use the time to consolidate some of its recent experiences while saving up some energy for effectively reporting future surprises.

The chapters that follow spotlight these theorized information pathways, as configured in various ways by psychology, neuroscience, and connectionism, in an attempt to further understand how brains predict, how they have evolved to do so, and how the neural mechanisms behind prediction might assist artificial intelligence researchers in building better, more adaptive, systems.

2 Conceptual Foundations of Prediction

Prediction is a catchy buzzword pertaining to speculation about the future, but many of its uses lack a temporal component and involve little more than basic causal inference. When my son comes home bloodied and bruised from his daily mountain-bike ride, I can predict that *the trail won today*. This hints of temporal lookahead, because my current state of knowledge, when he comes in the door, has not yet been updated to include the cause of his battered appearance. I am thus predicting what my son will soon tell me: he suffered multiple painful wipeouts. In short, my hypothesis about a past cause constitutes a prediction with respect to my future knowledge state. Subjectively, my causal hypothesis involves the future, not the past.

Similarly, in machine learning, a deep network trained to perform facial recognition of the citizens of Land-O-Plenty may be said to *predict* that the suspicious midnight visitor to a local ATM was Robin Green, based on a brief and blurry video sequence, when more accurate verbs include *estimate*, *speculate*, or *guess*, none of which suggest temporality. However the word *predict* might seem justified under the assumption that the authorities will eventually discern the true identity of the culprit, sometime in the future, thereby verifying or refuting my prediction.

In providing a basic conceptual backdrop for the remainder of the book, this chapter sticks as closely as possible to the definition of prediction found in *Webster's Dictionary*: "to declare or indicate in advance." However, there will always be the lingering question: In advance of *what*? Are we discussing the occurrence of an event on an objective timeline, or the awareness of that event by a particular agent, or the formation of a representation of that event by neural firings in a particular brain region of that agent? The philosophical slippery slopes are unavoidable, but I will do my best to avoid sliding too far down any of them.

In general, our deep investigation of prediction in neural systems will entail a rather subjective view. Signals can take hundreds of milliseconds to fully register in human sensory apparatus, and tens of milliseconds to travel between neurons. Hence, the prediction by a neural assembly of what it (or another assembly) will experience in the future will often be the consequence of some world event of the recent past (that the nervous system is gradually processing). So neurons predict what (possibly other) neurons will *see* or represent in the near future about a past event. Temporality is essential to all of this, but, unfortunately, often a bit confusing.

2.1 Compare and Err

My high school coach had a slogan taped to his dashboard: *Once I thought I was wrong, but I was mistaken.* Predictions are frequently mistaken, but the predictor need not be inexorably wrong. Typically, prediction is not a single act, but an adaptive process by which wrongs gradually get righted. In the terminology of control theory, predictors often operate in a closed-loop mode in which they make a prediction, compare it to a target to produce an error, and then use the error as the basis for an updated prediction, which then leads to another error and another prediction.

Letting P denote the prediction, E the error, and R the target / reality, a simple predictor algorithm for a perpetually active agent is

1. Initialize P

2. Input R

3. $E = R - P$

4. $P = \Omega(E, P)$

5. Go to 2

The function Ω may be as simple as $\Omega(x, y) = x + y$: it just adds the error to the predicted value to yield R, since $P + E = P + (R - P) = R$. That works fine as long as R changes very little between timesteps, which depends on the operational timescales of the environment and predicting agent. In complex situations, however, even for a reasonably static R, the intricacies of the agent's decision-making apparatus (e.g., a neural network) and of the agent-environment coupling may preclude the simple design of Ω or straightforward calculation of E. Both the comparison and the update may involve noise or other forms of stochasticity. Hence, many iterations through the loop may be necessary to gradually bring E's magnitude down to an acceptable level.

Some of these confounding factors and more advanced control loops appear later in this chapter and book. The main lessons for now are that (a) prediction is typically a closed-loop process, that (b) requires a comparison of the expectation to a target to produce an error, which (c) affects future predictions.

2.2 Guesses and Goals

Sticking as close as possible to the temporal characterization of prediction, it involves conjuring up (or at least behaving *as if* one has conjured up) the future state of some system. Our general conception of a prediction is thus a forecast (or guess) of a future state, whether in the next millisecond or the next century. However, few species *make a living* solely from predicting the future; they need to *act* in a manner that takes advantage of those predictions. Even among humans, those who earn money from forecasting do so because somebody exploits that information to better perform some activity, and is thus willing to pay for it. In short, a prediction isn't worth much if it doesn't enhance survival in some way or another.

Imagine a cheetah chasing a gazelle across the savanna. It may *predict* that the gazelle is trying to get to a wooded area, where it might have a better chance of evading its pursuer, but it will also *desire* to keep it in the open. Thus, the cheetah (implicitly or explicitly)

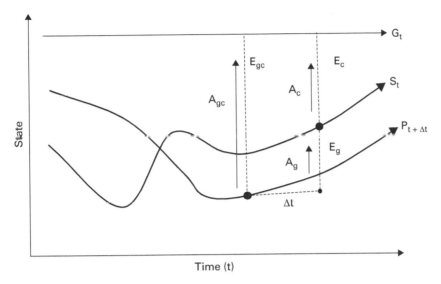

Figure 2.1
Dual aspects of prediction: guessing versus control. The three plots: S_t = state of system at time t, $P_{t+\Delta t}$ = prediction / guess at time t of the system state at time $t + \Delta t$, and G_t = goal state at time t. The simple errors pertain to control, E_c, and guessing, E_g, and the actions aimed at reducing those errors are A_c and A_g, respectively. The more complex error, E_{gc}, is between the goal and the predicted future state, while action A_{gc} incorporates that prediction in pursuit of the goal.

manages two alternate realities, *gazelle in woods* and *gazelle in wide-open space*, while dealing with the current state of the world. We can call these alternate realities the *guess* and *goal*, respectively. In their simplest form, all of these (world) states include two key elements: the locations of the cheetah and the gazelle; more elaborate versions would include the direction and magnitude of their velocities and accelerations, energy levels, signs of weakness, and so on.

As illustrated in figure 2.1, the goal state (G_t) can be assumed constant for the time frame of our analysis, while the current system state (S_t) changes, as does the guess of the future state $P_{t+\Delta t}$: at time t, the agent (cheetah) predicts the world state at time $t + \Delta t$. In the case of a pure forecasting problem, the agent's only concern is the reduction of the guessing error (E_g) via some sort of learning action (A_g): the agent tries to improve its forecast. In this mode, $S_{t+\Delta t}$ plays the role of the target value, and the agent modifies $P_{t+\Delta t}$ to try to match it.

Conversely, for a basic control problem, the agent seeks to change S_t to bring it closer to the target / goal G_t and thereby reduce the control error (E_c). The cheetah does so by acting in a manner that pushes the gazelle from its current location toward the wide-open savanna. However, cheetahs and gazelles move quickly, with the state of the body-world coupling changing too fast for the nervous system to keep up, so a successful cheetah will probably use its guessed state and that state's difference from the goal (E_{gc}) to govern its actions (A_{gc}).

Following the basic philosophy of relativism (that no truth or knowledge is absolute, only relative), an agent's goals represent desired states of the world *as perceived by the agent*. Thus, G_t constitutes a particular target state of the agent's sensory apparatus, essentially shrink-wrapping the agent's scope of perceived space and time onto its receptors. Goals thereby become very intimate states of the agent, as do predictions of future states. The simplest organisms lack explicit representations for both goals and predictions, though their actions give the impression (to an outside observer) that they have target states and maybe even guesses (as to how their prey will move in the next half second).

In moving up the ladder of cranial complexity, we find organisms able to create anticipatory states, but these guesses would probably confer a survival advantage only if they functioned as goals, and thus, $P_{t+\triangle t} = G_{t+\triangle t}$. A crab probably gains little from imagining future states per se, but by having some conception of a goal and how to nudge its current state toward that target, it would seem to have a claw up on any competitors who lacked the ability to represent alternative states. The next step, divorcing $P_{t+\triangle t}$ from $G_{t+\triangle t}$, surely took considerable evolutionary time. But the neural mechanisms that allowed the formation of one type of alternative reality, $G_{t+\triangle t}$, were probably usurped and enhanced to support the other, $P_{t+\triangle t}$.

Today, in analyzing the behavior of an artificial neural network that predicts the next value (V) in a sequence (e.g., tomorrow's high temperature based on the daily highs of the past month), we view V as the target, and the network's output as $P_{t+\triangle t}$: reality (V) is objective, and primary. But in a living agent, reality is subjective, and secondary to the agent's goals, toward which it will try to bend its perceived reality through actions on its own body and their influences on the surroundings. This egocentric view of behavior plays a vital role in understanding the rudiments of prediction in biological systems, and even some of the theoretical underpinnings of certain types of artificial neural networks.

In general, many discussions of prediction should begin by clarifying the viewpoint: egocentric (subjective) or ecocentric (based on a world or environment that embodies objectivity[1]). Thus, the neural networks studied in deep learning typically assume an ecocentric stance, with the data set being the objective truth, and the learning algorithm trying to modify the network parameters so as to generate outputs well-correlated with that reality. Conversely, agents (biological or artificial) that perform actions in a world (real or virtual) tend to have goals (such as survival and reproduction in living organisms) that take precedence over *building objective models of the world*. The models that they do craft are, more likely, biased by their own abilities to perceive and act. A pelican needs no detailed model of the ocean's depths to successfully dive for fish.

2.3 Gradients

Although change is ubiquitous in the dynamic environments that we inhabit, surprising (i.e., unpredicted) changes are much less frequent. The common (often correct) assumption that tomorrow will be much like today is not a belief in stasis so much as a prediction that tomorrow's events will *unfold* much like today's: the same basic pattern of change will repeat. This faith in reoccurring or gradually changing patterns (i.e., trends) underlies much of our predictive power. By monitoring and quantifying these trends, we build our foundation for speculation. And for the most part, we are reluctant to disregard these trends in the absence of unusual, unexpected sensory data.

Mathematically, these trends are often formalized as *derivatives* or *gradients*: the change in one variable ($\triangle y$) as correlated with or caused by the change in another variable ($\triangle x$). In other words, how *sensitive* is the value of y to changes in x? Figure 2.2 provides a simple graphic illustration of this concept, often described as *the rise over the run*.

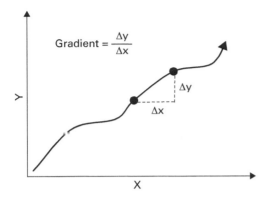

Figure 2.2
The basic mathematical concept of a gradient: the change in one variable ($\triangle y$) with respect to the change in another ($\triangle x$). Viewing y as a function of x, that is, y = f(x), then the gradient (or derivative) of y with respect to x is commonly written as $\frac{\partial y}{\partial x}$ or $f'(x)$.

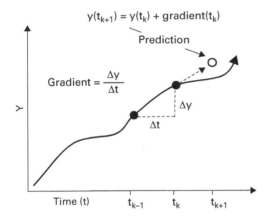

Figure 2.3
Gradients and their role in prediction. Basic interpolation using y's gradient (with respect to time) at time t_k enables a simple, primitive forecast of $y(t_{k+1})$.

The assumption of the repeated pattern plays out mathematically as an interpolation of a recent gradient to a point in the future. The graph of figure 2.3 uses time as the variable x and shows the gradient calculated from time t_{k-1} to time t_k, which is then extrapolated (dashed arrow) to a prediction for the value of y at time t_{k+1}. In this case, the short-term prediction is quite weak, since y briefly flattens out after time t_k, but the longer-term estimate looks more promising. Basically, the gradient serves as a poor man's tool for prediction, but one used frequently in both nature and technology.

Mathematically, the classic construct for prediction by gradients is the Taylor series, which provides estimates for $y_1 = f(x_1)$ when given $y_0 = f(x_0)$, $\triangle x = x_1 - x_0$, and all of the derivatives (first, f', second, f'', etc.) of f(x) at x_0:

$$f(x_1) = f(x_0) + f'(x_0)\triangle x + \frac{f''(x_0)}{2}\triangle x^2 + \frac{f'''(x_0)}{6}\triangle x^3 + \ldots \qquad (2.1)$$

Back in calculus class, the Taylor series was just another one of those oddities that we had to learn, but we probably never gave much thought to some inherent practical problems. After all, if we know f(x) and x_1, why can't we just plug x_1 into f(x) to produce y_1? Oh, but we don't really know f(x), yet we know a whole series of its derivatives at x_0? In fact, we can calculate an infinite sequence of those derivates at or near x_0 but we don't have a general understanding of f(x) itself. That is, we do not know f(x) *for each possible value of x*? What kind of world is this?

As it turns out, this is the real world outside of the calculus book, the world in which we have to make predictions based on weak information, because the general expression for some quantity y as a function of another quantity x eludes us. So we can measure y_0 at (some point in time or space) x_0, and we can measure how y changes as x deviates from x_0, and how the x-induced change in y changes as x changes, and how the x-induced change in the x-induced change of y changes as x changes, and so on. Then we can plug all of that information into the Taylor formula and estimate/predict y_1 at the new point x_1. This sounds great, but in so many cases, discerning anything beyond the first derivative becomes a real chore, and we're basically back to a fairly primitive interpolation similar to that of figure 2.3. But, again, that's often enough: many practical applications of the Taylor series only use the first derivative anyway.

2.3.1 Gradients Rising

Gradients arise in a wide variety of subject areas, and only in toy textbook problems of those domains is a general expression for f(x) actually known. Table 2.1 summarizes a few of these domains. Two key questions now surface in these and other areas: (1) What predictions do the gradients support, and (2) How do these predictions enable adaptive behavior?

In analyzing the foraging behavior of bacteria, biologists examine the changes in nutrient concentrations across space and compare those gradients to the movement patterns of the microorganisms. This provides a measure of *intelligence* in terms of whether or not bacteria swim up a nutrient gradient (i.e., in the direction of increasing nutrient) and down that of a toxic chemical. As evidenced by many bacteria's well-documented tendencies to (a) continue swimming along promising gradients, but (b) lapse into random movements in uninformative gradients, these organisms make implicit predictions as to the nutrients beyond their sensory horizons.

On the other end of the spectrum lie some of the more complex decisions made by any agent: stock trading. Here, one standard gradient is simply how the stock price has changed

Table 2.1
Diverse domains and variable pairs within each for which relevant gradients $\left(\frac{\Delta Y}{\Delta X}\right)$ play an important role.

Domain	X	Y
Bacterial Foraging	Location	Nutrients
Finance	Time	Stock Price
Thermoregulation	Heat	Temperature
Evolution	Genotype	Fitness
Brain Development	Location	Neurotrophins
Deep Learning	Connection Weights	Output Error

during the recent past, which can then be used to predict the future price, which, in turn, dictates trading actions, all in the service of maximizing profits. In temperature regulation, an engineered system uses several metrics, including the effects that changes in heat flow have recently had on ambient temperature; these indicate how future flow rates will affect temperature, which, in turn, determines how best to adjust flows in order to move temperature toward a target value.

Evolution (and similarly, evolutionary algorithms) displays interesting gradients, few of which directly assist in adaptive behavior but nonetheless provide enlightening perspectives on long-term population changes. For example, the sensitivity of fitness (F) to mutations of a particular portion of the genome (G) often indicates the degree to which G has evolved to satisfy environmental demands (or remains relatively decoupled from those selection pressures). If $\frac{\Delta F}{\Delta G}$ has high magnitude, G has probably been forced to bow to selection pressure over the generations and is also a likely target for genetic modifications that have significant phenotypic impacts. Such genes often show low variance in the population, since strong selection may favor a small subset of the alleles.

From a more theoretical perspective, evolution typically follows fitness gradients in an implicit manner. Consider the situation in figure 2.4, when two parents occupy different elevations on the fitness landscape. The more fit (higher) parent will tend to produce more offspring (that survive), and thus the population average will climb the gradient (without any explicit awareness, goals, or actions of the individuals or species). Gradient ascent is a simple, indirect, emergent property of evolution by natural selection.

The inset of figure 2.4 illustrates explicit gradient following in evolution, wherein parents (a) have full awareness of the fitness landscape, (b) can control the genomes of their

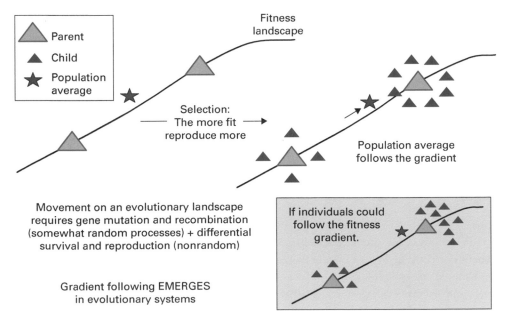

Figure 2.4
Evolution as gradient following on a fitness landscape. (Main diagrams) Implicit gradient ascension performed by evolution. (Inset) Hypothetical result of explicit gradient following (if parents could genetically engineer their children).

offspring, and (c) can predict phenotypic consequences of genetic change. This empowers parents to generate offspring above themselves on the fitness landscape, thus raising the population's average fitness. Notwithstanding contemporary genetic engineering, this is still mainly a hypothetical scenario; but the difference between the two images indicates the potential (disconcerting) acceleration of evolution incurred by designer babies.

Early development of the brain involves numerous spatiochemical gradients that perform functions such as laying out the body axes, providing pathways for the migration of neurons to their appropriate regions, and growing axons to proper dendritic targets. Few biological processes are as awe-inspiring as microscopic videos of the emergence of brains and bodies via the physical adaptations of cells based on the chemicals secreted by other cells.

Finally, in the field of deep learning (DL), gradients are ubiquitous. Their typical embodiment is in the derivative of a network's output error with respect to its tuneable parameters, for example, the weights on interneuronal connections. These networks learn by modifying the parameters to reduce the net's output error, in a manner governed by the gradients (which represent the change in error as a function of the changes in weights). A neural net for solving a complex problem may house millions of such gradients, most of which constitute *long-distance connections* in the computational structure of the network: many sequential calculations separate any of these parameters from the network's output (and hence from the output error term).

Each new DL variant typically includes creative accessory network components that introduce novel parameters into contemporary DL architectures, and gradients of output with respect to these new parameters must be calculated to enable learning. The key to implementing any new architecture lies in mathematically formalizing the complex gradients.[2] These long gradients can yield impressive results on machine learning tasks, often outperforming humans, but they have little biological realism and therefore have limited utility as models of natural cognition.

Many of the artificial neural networks (ANNs) covered in this book exhibit different dynamics than the DL architectures currently in vogue. They are predecessors to some of today's networks, but with different behavioral goals, formalized as objective functions with properties inherited from thermodynamics. These objective functions are mathematically shown to decompose the biologically implausible long derivatives into local gradients (of prediction error with respect to nearby synaptic weights). Crucially, these prediction errors are local to each neural layer as components of a hierarchical brain model known as *predictive coding* (Rao and Ballard 1999), a central topic of this book.

2.4 Sequences

Prediction tasks are often couched as sequence-completion problems, such as $1,4,7,\Theta$, where correctly finding Θ entails making an accurate prediction. This could be phrased as a simple analogy problem: *1 is to 4 as 4 is to 7 as 7 is to* Θ. The problem becomes trivial once one discovers the relationship between any two adjacent elements and then applies that relationship to 7 to yield Θ's value, 10.

Note that with number sequences, the relationship constitutes a gradient, $\frac{\Delta N}{\Delta P}$, where N denotes the sequence element and P is the position index. In this case, simple observation reveals that $\frac{\Delta N}{\Delta P} = 3$, and the prediction is $\Theta = 7 + \frac{\Delta N}{\Delta P} = 7 + 3 = 10$.

This easily scales up from single numbers to points in a multidimensional space. Can you solve for Θ in the following sequence?

$$[1, 5, 8], [3, 3, 18], [5, 1, 28], \Theta \tag{2.2}$$

In this case, we compute gradients (relationships) in each dimension independently, then apply them to [5,1,28] to produce Θ. These gradients are $[2, -2, 10]$, so adding this to [5,1,28] yields $\Theta = [7, -1, 38]$, and we expect all four of these points to lie along a straight line in three-dimensional space.

When we go *off the (Cartesian) grid* into a nonnumeric space, the problems become slightly harder, for example: A, D, G, Θ. Solve for Θ. Although most people tackle this problem with ease, they do so despite a host of potentially confounding issues involving gradients: What is the salient relationship that links A to D and D to G? Is it the difference in sounds made by a speaker of language L when pronouncing each letter, or maybe the sound frequencies produced by a piano when playing each of these notes? How about a speaker of language M or a saxophone player? We need to establish an underlying substrate, known in mathematics as a *metric space*, in which *distance* has a definition similar to our normal conception from Cartesian space.

In the first sequence, we simply mapped the numeric symbols 1, 4, and 7 to a number line in one-dimensional space, probably without even thinking about it. And, almost as fluidly, most people would map A, D, and G to their indices in the English alphabet (again, 1, 4, and 7), which then map to the number line. Once in the metric space, we can compute the gradient (3), add it to our rightmost[3] point (7) to produce 10. As a final step, we then project 10 back into English-alphabet space, where the tenth letter is J. So $\Theta = J$.

What if the sequences involve words, such as *jot, lot, not,* Θ? In this particular case, each word readily maps to three indices (one for each letter) in Cartesian space: [10,15,20], [12,15,20], and [14,15,20], and each adjacent pair has the gradient vector [2,0,0]. Adding this to the vector of *not*, gives [16,15,20], which projects to the word *pot*. So Θ = pot.

Of course, most word sequences require different tactics. If they represent familiar phrases, then previous exposure allows pattern completion by rote memory. For example, Elvis Presley fans can easily solve for α and β in *You, ain't, nothing, but, a,* α, β. But the words *hound* (α) and *dog* (β) do not naturally follow from the others in any obvious Cartesian space. Neither do *heart* (α) and *break* (β), the likely prediction from fans of Backstreet Boys. Eager listeners of both artists might request more context (i.e., a few more predecessor words) to disambiguate the two options, but these would only enhance the contextual priming of rote memory while decreasing the (already remote) possibility of finding some useful Cartesian space in which gradients could drive prediction. When it comes to predictions involving overly familiar sequences, the only Cartesian space may be the one expressing the temporal order in which you (repeatedly) heard the phrase, but not even that would provide a meaningful gradient for predicting the completion. Basically, rote memory requires no real understanding of the sequence elements, just a mental connection between them based on nothing more than their temporal juxtaposition.

Somewhere between [1,4,7,Θ] and complete-that-tune lie prediction problems that require deep understanding of the situation but suggest no obvious Cartesian spaces in which gradients can make a contribution. However, finding useful metric spaces may require only a bit of creativity and general experience in the world. Consider this sequence: white bear, brown gopher, green snake, Θ. If you cannot embed this sequence in a fairy tale or nature program to give it some useful semantics, a decomposition into two metric spaces may prove useful: size and seasonal colors. Since bears are larger than gophers, which are larger than (most) snakes, the prediction that Θ is *smaller than a snake* makes sense. In many parts of the world, each season has a dominant color, with the first season (winter in the northern hemisphere) characterized as white, the second (early spring) as brown, then green in the summer, followed by the oranges and reds of autumn. Thus, an intelligent guess for Θ might be *orange beetle*.

In fact, true understanding of a concept (even one as intangible as *disappointment* or *omnipresent*) may demand its embedding in a metric space in which gradients have predictive power. As radical as this may sound, it seems fairly straightforward to place most concepts on some form of spectrum involving several others, such as [devastated, depressed, melancholy, disappointed, bored, complacent, content, happy, gleeful, elated]. We can then compute gradients of this space with respect to another factor, such as the amount of positive reward received by an agent. Thus, any increase (decrease) in reward should cause movement in the direction of elation (devastation). Understanding arises from these embeddings, probably several for any given concept. By relating them to one another and calibrating how both external factors and one's own actions can move along these spectra, an agent gains salient knowledge and survival advantages above and beyond those accrued by knowing all Elvis Presley' lyrics by heart.

This, very brief, foray into sequences and metric spaces and their relationships to understanding and prediction scurries around the rim of a very deep philosophical crater that was fearlessly explored by Peter Gärdenfors in his classic work, *Conceptual Spaces* (2000). The main purpose of this section is only to provide a rudimentary grounding of what many consider the classic prediction problem, sequence completion, in spaces where distance and gradients make sense. In chapter 3, when we look at the hippocampus, the grounding will descend further, to the level of individual neurons.

2.5   Abstracting by Averaging

Predictions based on gradients put faith in recent (and/or generally understood) changes and their continued relevance: yesterday's run on sirloin steaks at the local supermarket will continue today. Obviously, this is not a foolproof strategy, for example, if yesterday was the final day of the holiday barbecue season. When quantities fluctuate, that is, exhibit positive and negative temporal gradients in a short time span, then the safer approach is to simply compute a scaled sum (aka average) of the values during that period and use it as the prediction. For example, given little more information than the past month of temperatures, a reasonable guess of tomorrow's temperature is the monthly average, particularly in those equatorial parts of the world having little seasonal variation.

The sum (or integral, for continuous systems) provides a very basic, but irreplaceable, tool for prediction. Unlike the gradient, which essentially adds a level of nuance, the average elevates decision making to a higher level of abstraction: one that glosses over the details in favor of the scaled conglomerate, accumulated over space and/or time. Predictions based primarily on sums may totally fail in the short term—yesterday's sudden uncharacteristic drop in a stock price may continue, unabated into today's trading—but still provide stellar long-term performance: the stock proves stable over the entire summer and provides a safe haven for calm investors.

Along with the conventional arithmetic average (the sum of a set of values divided by the size of the set), the weighted average plays a very important role in prediction. In the equation below, Θ represents the weighted average of the x_k values:

$$\Theta = \sum_{k=1}^{N} w_k x_k \text{ where } \sum_{k=1}^{N} w_k = 1$$

For the standard average, all weights (w_k) are identical and equal to $\frac{1}{N}$, but in many predictive situations, the x_k values closest (in space or time) to the value to be predicted (Θ) will have more relevance and thus have higher weights than the more distant values. For instance, when predicting the grade-point average (GPA) of next year's incoming class of mathematics students, the department administrator might average over the last ten years of (average) GPAs, but with the highest weights given to the past two or three years.[4] Weighted averages of this sort are ubiquitous in AI algorithms, particularly those involving a time component, such as reinforcement learning (RL). They also loom large in spatial domains, such as image processing, where filters applied to small regions of an image often exhibit biases toward nearby pixel values.

This book examines prediction in hierarchical neural systems, where higher levels produce expectations of values at lower levels. These upper predictors typically run at slower timescales. Thus, they are less jittery, and tend to react to averages of lower-level behavior more than to momentary changes, although they may, in addition, accumulate those changes into an average gradient. Regardless, the inherent delays in the system make it infeasible for these slower responders to make predictions solely on the basis of a single gradient (which may be outdated / *stale* by the time it is received). Thus, the higher levels are forced to rely heavily on quantities that have been abstracted over time and/or space, with the typical abstraction tool being scaled summations akin to weighted averages.

2.6 Control and Prediction

Armed with gradients and weighted averages, we can revisit the basic control problem of figure 2.1 and formulate a simple predictive scheme. Figure 2.5 depicts the same situation as earlier, but now with the focus on calculation of the *guess*, $P_{t*+\Delta t}$. Define two positive factors, k_g and k_a, for the gradient and average, respectively.[5] Thus, the prediction becomes a weighted combination of gradient and biased average. The gradient, based on timestep Δt, goes back one time increment, while the weighted average covers the present state value (S_{t*}) plus four historical points (drawn as gray octagons).

One reasonable formula for the prediction is then

$$P_{t*+\Delta t} = \underbrace{k_g \left(S_{t*} + \frac{\Delta S_{t*}}{\Delta t} \right)}_{\text{gradient-based}} + \underbrace{k_a \left(\frac{9}{16} S_{t*} + \sum_{j=1}^{4} \frac{1}{2^{j+1}} S_{t*-j\Delta t} \right)}_{\text{average-based}} \qquad (2.3)$$

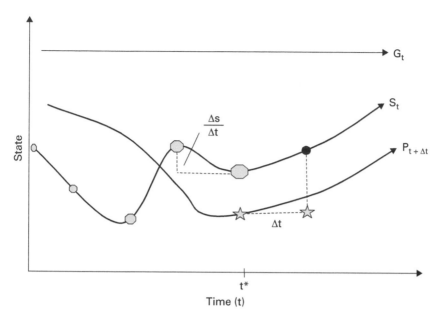

Figure 2.5
Basing prediction on a combination of gradient and average, using the same scenario as depicted in figure 2.1. At time point t*, to compute the prediction ($P_{t*+\triangle t}$, denoted by the gray star) for state S at time $t* + \triangle t$, combine the gradient $\left(\frac{\triangle S}{\triangle t}\right)$ and a weighted average over S values in the recent past (gray octagons). Shrinking size of octagons going backward in time represents decreased weighting of those points in a biased average.

where the first average term's weight $\left(\frac{1}{2} + \frac{1}{2^4} = \frac{9}{16}\right)$ is assigned such that all weights sum to 1. Equation 2.4 gives a more generic form, averaging over M+1 points, $\vec{S}_t$, with weights, w_j, that sum to 1.

$$P_{t+\triangle t} = \Gamma(\vec{S}_t) = \underbrace{k_g S_t + k_g \frac{\triangle S_t}{\triangle t}}_{gradient\text{-}based} + \underbrace{k_a \sum_{j=0}^{M} w_j S_{t-j\triangle t}}_{average\text{-}based} \qquad (2.4)$$

In a typical control problem, the error, $E_t = G_t - S_t$, governs the choice of a control output, commonly designated as u_t, which then feeds into the system being controlled (aka the *plant*), which produces a new state $S_{t+\triangle t}$. Letting Φ denote the controller, and Ω the system / plant:

$$E_t = G_t - S_t \qquad (2.5)$$

$$u_t = \Phi(E_t) \qquad (2.6)$$

$$S_{t+\triangle t} = \Omega(u_t, S_t) \qquad (2.7)$$

For simple systems, Φ may just multiply the error by a positive fractional constant to try to bring S closer to G. However, as discussed earlier, the delays inherent in the sensorimotor systems of most living organisms preclude the use of the current state as the basis for action. Rather, a prediction of the future state (at time $t + \triangle t$) is necessary, and it, in turn, will yield a predicted future error, which can then initiate an action, u_t, which will not take effect until time $t + \Delta t$, due to motor delays. Considering all of the different delays will unecessarily complicate the analysis, but a simple update of the model yields

$$P_{t+\Delta t} = \Gamma(\vec{S_t}) \tag{2.8}$$

$$E_{t+\Delta t} = G_{t+\Delta t} - P_{t+\Delta t} \tag{2.9}$$

$$u_t = \Phi(E_{t+\Delta t}) \tag{2.10}$$

$$\vec{S}_{t+\Delta t} = \Omega(u_t, \vec{S_t}) \tag{2.11}$$

$$P_{t+2\Delta t} = \Gamma(\vec{S_{t+\Delta t}}) \tag{2.12}$$

In this model, note that the action choice for time t (u_t) is a function of the estimated error at $t + \Delta t$. If Φ represents a standard proportional-integral-derivative (PID) controller, it computes the control output, u, as a function of the current value, derivative, and weighted average of the error term, not of the state itself and not of a single value of the error, but of the whole history of errors. Assuming that G_t, the goal, is a constant (G) throughout the regulatory period, the expression for error is simply $E_t = G - S_t$. When the situation demands a predicted error, $E_{t+\Delta t}$ as input to Φ, it can be derived from the current and previous errors via an identical formula to equation 2.4, but with $\vec{S_t}$ replaced by $G \overset{\rightarrow}{-} S_t$:

$$E_{t+\Delta t} = \Gamma(G \overset{\rightarrow}{-} S_t) = \underbrace{k_g(G - S_t) + k_g \frac{\Delta(G - S_t)}{\Delta t}}_{\text{gradient-based}} + \underbrace{k_a \sum_{j=0}^{M} w_j(G - S_{t-j\Delta t})}_{\text{average-based}} \tag{2.13}$$

Compare this to a standard discrete model of the PID controller, wherein the control output, u_t, stems directly from the error terms:

$$u_t = k_p e_t + k_d \frac{\Delta e_t}{\Delta t} + k_i \sum_{j=0}^{t} e_j \tag{2.14}$$

where e_t is the goal state minus the current state.

The similarities between equations 2.13 and 2.14 are obvious, and none of the differences detract from the main conclusion: making predictions and controlling a system are very similar operations. This becomes clear after a few more manipulations.

Since G is constant and $\sum_j w_j = 1$, the following simplifications hold:

$$k_g \frac{\Delta(G - S_t)}{\Delta t} = -k_g \frac{\Delta S_t}{\Delta t} \tag{2.15}$$

$$k_a \sum_{j=0}^{M} w_j(G - S_{t-j\Delta t}) = k_a G - k_a \sum_{j=0}^{M} w_j S_{t-j\Delta t} \tag{2.16}$$

Putting this all together yields

$$E_{t+\Delta t} = (k_a + k_g)G - k_g S_t - k_g \frac{\Delta S_t}{\Delta t} - k_a \sum_{j=0}^{M} w_j S_{t-j\Delta t} \tag{2.17}$$

By the expression for the predicted state in equation 2.4,

$$E_{t+\triangle t} = (k_a + k_g)G - P_{t+\triangle t} \qquad (2.18)$$

Hence the predicted error is simply a constant minus the predicted state, and computing a typical PID control output is akin to computing the predicted state: predicted values, prediction errors, and control decisions are nearly proxies for one another, with the same pivotal computations (using the same gradients and averages of system states) underlying all three. Very little separates prediction from control, and from a biological perspective, the ability to predict may have arisen from an organism's fundamental need to control, both its internal and external environment.

Control theory is no stranger to neuroscience, particularly in studies of the cerebellum (Wolpert, Miall, and Kawato 1998). More generally, brains function as the central controller of myriad physiological processes, from respiration, circulation, and waste removal to hormone balancing and growth. Although descriptions of these homeostatic processes rarely invoke prediction, that story has begun to change, as indicated by these introductory lines from neuroscientists Peter Sterling and Simon Laughlin in *Principles of Neural Design* (2015, xvi):

... the core task of all brains: It is to regulate the organism's internal milieu—by responding to needs and, better still, by *anticipating needs* [my emphasis] and preparing to satisfy them before they arise. The advantages of omniscience encourage omnipresence. Brains tend to become universal devices that tune all internal parameters to improve overall stability and economy. *Anticipatory regulation* replaces the more familiar *homeostatic regulation*—which is supposed to operate by waiting for each parameter to deviate from a *set point*, then detecting the error and correcting it by feedback.

They go on to extoll the advantages of anticipation in sympathetic and parasympathetic regulation, where many of the problems that prediction ameliorates (such as internal load imbalances and supply-demand mismatches) stem from the time lags of corporeal homeostasis, just as latencies in the sensorimotor system create problems for a pure sense-and-react agent. Anticipatory regulation can also spill over into overt behavior, as animals search for particular environmental niches that will serve future metabolic or reproductive needs, for example, a watering hole before the actual onset of thirst, or a secure nest site prior to mating season. The next step in this progression is to behaviors that anticipate changes in the environment or body-world coupling, that is, the more common, ethological notion of prediction.

2.7   Predictive Coding

The notion of *predictive coding* stems from neuroscientific work in the 1980s and back to psychological theories and engineering methods of the 1950s and 1960s. This section introduces the main aspects, which arise in many contexts throughout the book, while chapter 5 covers the topic in depth.

Imagine a large hierarchical organization such as a private corporation, national military, or university. In each, information flows in at least two directions: top-down and bottom-up. Although every member appreciates being informed, few function well under data overload conditions. A good deal of the hierarchy's effectiveness stems from an efficient handling of the signal flow: getting people the information that they need, but not burdening them with irrelevancies and redundancies. Thus, a middle manager receives updates from many lower-level workers but creates an executive summary as her main message to the upper eschelons, who then receive only the *vital essence* of the current situation, not the 1,001 individual stories and complaints. Similarly, many of the requests by individual workers are handled by managers, without relaying so much of the problem or solution upward. Most organizations work best when the vast majority of problems can be handled locally, without involving large chunks of human time and energy.

Similarly, when commands flow downward from the big brass to the managers and workers, the most efficient outcome is when operations get carried out to the leaders' satisfaction: when their expectations are met. A leader's favorite feedback from the level below is often, "Done," without a lot of elaboration. This normally signals that the predicted outcome has been achieved: mission accomplished.

Conversely, when problems arise, when visions / predictions do not come to fruition, the need for greater information rises steeply. Leaders and managers require more details (feedback) in order to adjust their expectations and solution strategies before sending updated commands back down the hierarchy. As the organization as a whole converges on a solution, as the expectations near realization, the signal flow can return to a trickle of simple commands, thumbs-ups and high fives.

This view of solving problems hierarchically while minimizing information transmission motivates predictive coding, which abstracts and formalizes the basic idea of two-way information flow and signal filtering. In the standard conception, predictions / expectations move top-down and sensory / reality signals travel bottom-up, and all aspects of reality that match coincident predictions can be removed (as redundant) from the upward signal: only mismatches between expectations and reality deserve further attention at higher levels.

This is illustrated in figure 2.6, where predictive coding's bottom-up *reality* signals serve as *target* values for top-down *predictive* signals. A comparator subtracts the prediction from the target to yield the *prediction error*, which then gets sent further upward in the neural hierarchy. Ideally, these bottom-up error signals attenuate up the hierarchy, with each prediction removing more of the residual, unpredicted target signal. Essentially, each level attempts to *explain away* targets at the next-lower level via its predictions.

The key implication of this model is that only unpredicted signals, that is, *surprise*, need travel very far up the hierarchy, thus saving the energy of transmitting a lot of redundant information. In short, if certain sensory information is expected, then why should a brain waste resources sending it around? The expectation alone, when unviolated, should be sufficient to trigger other activities, such as the proper motor responses.

Until we consider more details of predictive coding, the main aspects to keep in mind are the reality-prediction comparison and the fact that the resulting prediction error constitutes the primary signal upward in the hierarchy, the executive summary containing all and only the information needed at the next level.

2.8   Tracking Marr's Tiers

In his seminal work in the early days of computational neuroscience, David Marr (1982) introduced a three-tiered framework for analyzing complex systems such as brains and AI systems: computational, algorithmic, and implementational. These names, though

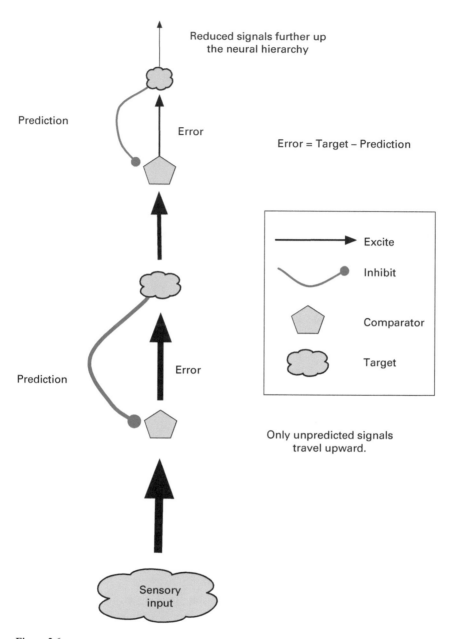

Figure 2.6
A simple illustration of predictive coding, with target signals moving upward in the neural hierarchy while gradually being reduced by predictions. Connection thickness mirrors the strength of upward and downward signals in this ideal situation of bottom-up error reduction.

somewhat misleading, correspond to three basic questions:

1. *What* is the basic phenomena to be modeled or problem to be solved?
2. *How*, at the algorithmic or pseudocode level, do you describe the plan of attack?
3. *Which* substrate will actually perform the computations? Transistors, DNA molecules, neurons?

So far, our discussions at the computational level have concerned one primary phenomenon: prediction, though others such as adaptation, emergence, and control play important roles as well. The current chapter has moved to the algorithmic level by showing how predictive tasks flesh out in terms of very basic computations: weighted sums, differences, products, and quotients.

Those predictions that lack this basic computability (via mental movement in a spatiotemporal or conceptual space) may rely on rote associative learning. For example, in classical conditioning experiments that link some random conditioned stimulus (e.g., a flashing light) to an unconditioned stimulus (e.g., an electric shock), the light allows the animal to predict the shock and then act to avoid it. Or, on more pleasant notes, song learning facilitates predictions of successive words that typically have no analogous neighbor relationship in a metric space. Sums, products, and gradients might help a point guard hit a cutting forward with a perfect pass,[6] but they won't help her sing the national anthem.

All of this sets the stage for a journey down to the implementational level of neurons and neural networks, where we have no problem finding neural mechanisms that *can* carry out the primitive operations of prediction in artificial nets and *might* be doing so in real brains as well.

3 Biological Foundations of Prediction

Adaptive behavior governed by predictions begins at the very lowest level of life and continues up through the phylogenetic tree. Gradient detection plays a vital role in survival and arises via pure biochemistry in single-celled organisms and then via neural circuitry in some of the simpler multicellular lifeforms. Many of the same neural principles then enable differentiation and prediction in the brains of higher animals.

This chapter investigates several of these primitive mechanisms, starting with the biochemical and quickly moving up through the neurophysiological to our ultimate goal, the neuroarchitectural. At that level, a diverse collection of neural circuits appears to implement prediction, and in many different ways, but all of which have direct ties to the general concepts introduced in chapter 2. Noticeably absent from this chapter is one dominant predictive region, the neocortex, whose detailed investigation is saved for later chapters on predictive coding and brain evolution, where the building-block modularity of the cortex meshes naturally with both traditional predictive-coding implementations and theories of incremental cognitive emergence.

3.1 Gradient-Following Bacteria

E. coli and other bacteria exploit a simple but effective strategy for moving along nutrient gradients. As shown in figure 3.1, when moving in the direction of increasing resource, the individual maintains a relatively straight line. However, when devoid of a promising gradient, it reverts to a random *tumbling* motion that basically amounts to exploration.

This simple strategy still requires a sophisticated physicochemical process to link chemical gradients to action selection. This begins by using a temporal derivative as a proxy for a spatial derivative: proximal sensory readings of an attractant (A) taken by an agent moving from point s_k (at time t_k) to point s_{k+1} (at time t_{k+1}) allow it to estimate $\frac{\Delta A}{\Delta t}$, which clearly mirrors $\frac{\Delta A}{\Delta s}$ (Dusenbery 1992). Intuitively, if moving in a straight line and detecting an increase in attractant over time, the agent has also recognized an increase across space.

As is common in biological systems, the actual mechanisms do not map cleanly to the computational logic followed by, for example, a gradient-following robot. Rather, the standard engineering variables and calculations exist only implicitly in the natural system. In the case of bacteria, no internal variable explicitly represents $\frac{\Delta A}{\Delta t}$, but the causal relationship between sensors and actuators yields overt gradient-following behavior.

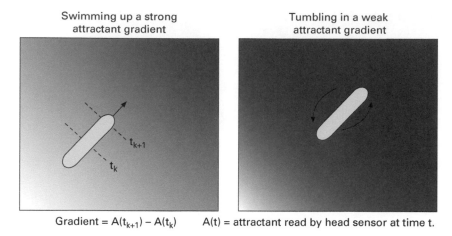

Figure 3.1
Basic swimming and tumbling behaviors employed by gradient-following bacteria (Levit and Stock 2002; Dusenbery 1992). Changes in background shading indicate gradients of chemical attractant.

Figure 3.2 illustrates the basic network of interactions operating within and around the *E. coli*. Both attractants and repellents bind to cell-membrane receptors, with opposite effects on internal kinase activity, which disrupts the coordinated movement of the cell's flagella, thus causing a tumbling motion. In low-kinase situations, the flagella rotate in the same direction and the cell swims in a constant direction. The swimming and tumbling motions constitute an opposing pair. As the diagram shows, attractant inhibits kinase and should thus reduce tumbling and disinhibit straight-line swimming.

However, gradient detection requires additional chemical dynamics. When attractant binds to the receptor, it also promotes the receptor's methylation, yielding it less sensitive to the attractant. This desensitization ensures that continued kinase deactivation will require an increased attractant concentration to bind the more finicky receptor. Thus, keeping kinase and tumbling in check requires an ever-increasing level of attractant.

The relationships among attractant, kinase, and movement appear in the rough sketch of figure 3.3, where (on the left) a relatively stable attractant level loses its inhibitory effects on kinase, and the dominant motion switches from swimming to tumbling. On the right, an ever-increasing attractant concentration holds kinase levels down such that swimming remains dominant.

In summary, these bacteria link ambient chemical concentrations to motion via an opposing pair of flagellar controls mediated by kinase; but tying chemical *gradients* to action requires the additional process of desensitization, which they achieve by receptor methylation. As it turns out, similar mechanisms operate in some of the neural circuits that appear to compute derivatives.

3.2   Neural Motifs for Gradient Calculation

As elaborated by Tripp and Eliasmith (2010), many types of neural circuits can implement gradient detection. Figure 3.4 portrays one of these models, wherein a detector neuron (A) for some quantity (e.g., an attractant) signals another neuron (B), which habituates to A's

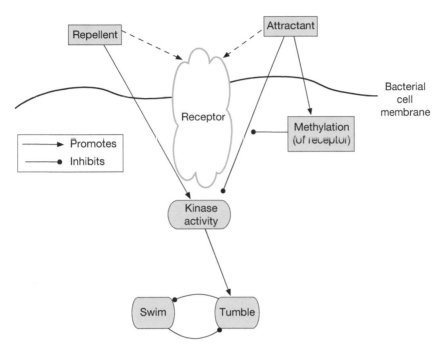

Figure 3.2
Chemical circuits for chemotaxis in *E. coli* bacteria, as described by Levit and Stock (2002).

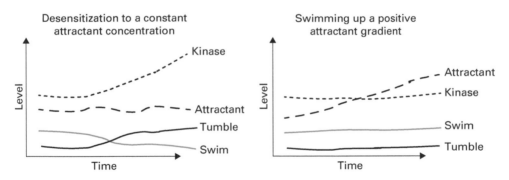

Figure 3.3
Sketch of the general progression of chemical concentrations and activities (swim, tumble) underlying bacterial chemotaxis, based on descriptions in Dusenbery (1992) and Levit and Stock (2002).

input, thus requiring an increase in A's activity for continued firing of B. Detailed analysis of the nematode worm *C. elegans* reveals this model in action (Larsch et al. 2015), albeit with considerable biochemical complexity. The relevant portion of the worm's nervous system, sketched in figure 3.5, shows many connections between a set of sensory neurons and a set of interneurons. Using this network, *C. elegans* can detect chemical gradients that span a concentration range of five orders of magnitude. Two key neurons in this process are AWA and AIA. The former desensitizes to the chemical attractant, while the latter desensitizes to signals from AWA. That combination enables the nematode to swim up very weak (yet nonzero) concentration gradients. In addition, the link from AWC (which is inhibited by the

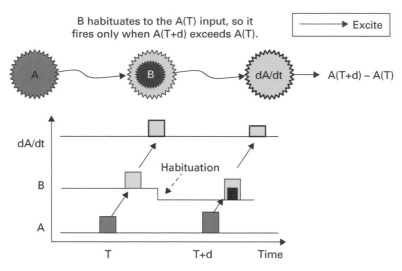

Figure 3.4
A simple neural circuit for detecting a positive derivative of a stimulus. Neuron A responds proportionally to stimulus concentration and excites neuron B, which desensitizes to A's input (indicated by the dark inner circle), thus requiring stronger signals from A in order to fire. The graph displays the temporal progression of this behavior, with the dark inner box on B's plot indicating an elevated firing threshold. The third neuron simply detects B's output and implicitly represents $\frac{\Delta A}{\Delta t}$.

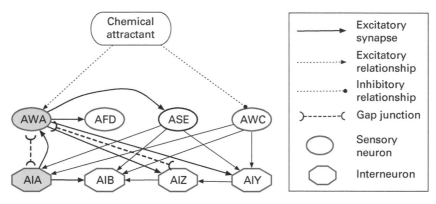

Figure 3.5
A portion of the circuit diagram of the nematode worm *C. elegans*, based on diagrams and descriptions in (Larsch et al. 2015).

chemical attractant) to AIA and other interneurons appears to mediate responses to weakly decreasing gradients, once again via desensitization.

As mentioned above, for simple organisms, one key functional consequence of desensitization is the extended *range* of stimuli that sensory receptors can detect. The mammalian visual system also exploits this relatively simple trick at multiple levels of neural processing to perceive a ten-thousand-fold range of light intensity, despite the fact that standard principles of neural coding admit only approximately a hundred unique neural signals (Dunn, Lankheet, and Rieke 2007). Thus, desensitization allows a sensory system to contextualize inputs such that *contrast* (in time and/or space) carries more salient information than do

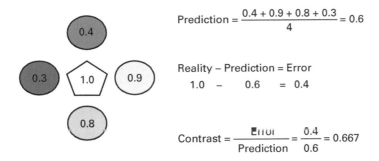

$$\text{Prediction} = \frac{0.4 + 0.9 + 0.8 + 0.3}{4} = 0.6$$

Reality – Prediction = Error
1.0   –   0.6   = 0.4

$$\text{Contrast} = \frac{\text{Error}}{\text{Prediction}} = \frac{0.4}{0.6} = 0.667$$

Figure 3.6
A central cell (pentagon) and its neighborhood (circles), with numbers denoting luminance in the range 0 (dark) to 1 (bright). Accompanying equations show basic relationships (as further discussed in the text) among the neighborhood average (a prediction of X's value), the difference between X's actual value and that average, and contrast.

absolute stimulus levels; and contrast, a concept with many definitions, clearly represents a gradient in space and/or time.

As portrayed in figure 3.6, contrast ties directly to prediction in common interpretations of neural processing, particularly in the early sensory layers (Sterling and Laughlin 2015). Essentially, the average level of stimulus among neighbors to unit X constitutes a prediction of X's own stimulus level. The difference between X's actual value and this prediction then yields an error term, and scaling that error by the prediction gives a value known as the *Weber contrast*. As discussed more thoroughly in later chapters, this manifests *predictive coding* (Rao and Ballard 1999), which saves considerable resources associated with signal transmission by requiring that only scaled errors (which embody *surprise*) be sent upward in the neural hierarchy. Hence, stimuli that match predictions (and are thus expected), incur very little signaling cost.

This is just one of many incarnations of prediction in the brain. In this case, the relationship between gradient (i.e., contrast) and prediction is reversed in that the former is derived from the latter, and the former becomes the key piece of information sent between neural layers. However, when sent upward, this gradient signal contributes to neighborhood-based predictions in the next level of the neural hierarchy. Contrast exemplifies the *differences that make a difference* touted by Edelman and Tononi (2000) as key elements of neural information processing, and it plays a key role in predictive coding.

Figure 3.7 indicates another mechanism by which neural circuits can detect derivatives (Sterling and Laughlin 2015; Tripp and Eliasmith 2010), in this case via the combination of delay routes and signal inversion. Imagine a simple example involving a series of sensory signals: 3,5,10,7,0,0 received at neuron A from time 1 to time 6. Assume that (a) neurons A and B produce an output signal in direct proportion to their input, but (b) neuron B's signal has an inhibitory effect on its downstream neighbor, and (c) passage through each neuron takes approximately the same nonzero amount of time (d in the diagram). As seen in table 3.1, the network's rightmost output equals the differences between successive inputs, that is, an estimate of the temporal derivative.

Chaining these simple modules together facilitates the computation of higher-order derivatives, as shown in figure 3.8. A few simple modifications to the lower circuit of figure 3.8 produces that of figure 3.9, in which an increase (from 1 to 2) of the weights (w)

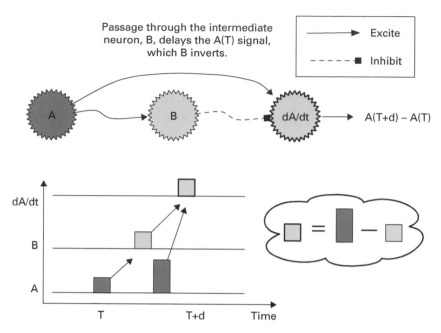

Figure 3.7
Illustration of temporal derivatives computed via a delay pathway and an inverted signal.

Table 3.1
Time series of each neuron's input(s) and rightmost output for the
network of figure 3.7 when given input sequence 3,5,10,7,0,0,
with the processing time of each neuron (d) being 1.

Time	A	B	$\frac{\partial A}{\partial t}$	Output
1	3	0	[0, 0]	0
2	5	3	[3, 0]	0
3	10	5	[5, −3]	0
4	7	10	[10, −5]	2
5	0	7	[7, −10]	5
6	0	0	[0, −7]	−3

on the leapfrogging connections changes the computed value from a derivative to the previous value plus the derivative: a primitive prediction of the future value $(x + \triangle x)$. Thus, by
sending downstream its value (weighted by w) and a delayed, inverted version, that neuron's
level receives a feedback prediction of its next value.

A few basic simulations, using a set of difference equations that directly model the
behavior of the lower circuit of figure 3.8, verify that these networks do indeed produce
higher-order derivatives. First, let the generator of input values be a sine curve: $y = sin(kt)$
where t is the integer timestep and k = 0.25. From basic calculus, it follows that $\frac{\partial y}{\partial t} =$
$kcos(kt)$, $\frac{\partial^2 y}{\partial t^2} = -k^2 sin(kt)$, and $\frac{\partial^3 y}{\partial t^3} = -k^3 cos(kt)$. Thus, the k+1st derivative is shifted one
quarter phase relative to the kth derivative and has one-fourth the amplitude. The plots of
figure 3.10 (left) clearly show the proper shifting and scaling of the outputs produced by the

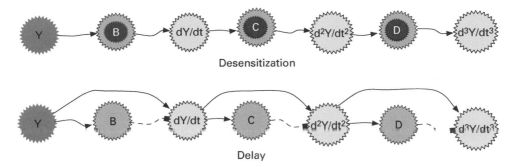

Figure 3.8
Neural circuits for computing three levels of derivatives using desensitization (top) and delayed inhibition (bottom).
All connections have weights 1 (excitors) and −1 (inhibitors).

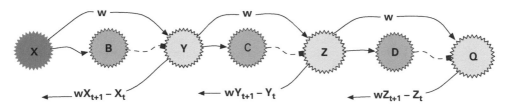

Figure 3.9
A simple neural circuit in which the proper choice of weighting ($w = 2$) enables the next value of neurons X, Y, and Z to be predicted by layers Y, Z, and Q, respectively. For example, at time 2, Y computes $2X_1 - X_0 = X_1 + (X_1 - X_0) = X_1 + \triangle X_1$. As in figure 3.8 (bottom), neurons B, C, and D act as delayed inverters.

network's three derivative neurons: the network has properly differentiated the input to the third order and could easily continue higher if given more neurons.

As a second example, consider a simple polynomial: $y = \left(\frac{t}{10}\right)^3$. Then, $\frac{\partial y}{\partial t} = \frac{3}{10}\left(\frac{t}{10}\right)^2$, $\frac{\partial^2 y}{\partial t^2} = \frac{6}{100}\left(\frac{t}{10}\right)$, and $\frac{\partial^3 y}{\partial t^3} = \frac{6}{1000}$. As seen in figure 3.10 (right), the delayed-inhibition network accurately computes three orders of derivatives.

In figure 3.10, note that the curves get flatter with the higher-order derivatives, until the third derivative is constant (or nearly so). In other words, at each higher level of the derivative hierarchy, the values change more slowly over time. This is common with polynomials, logarithms, and even exponentials (with fractional exponents such as $y = e^{t/5}$). Thus, for any neural network that implements this derivative hierarchy, the time constants at the higher levels can be higher (i.e., these layers can update more slowly) while still computing accurately.

In fact, the higher-order derivatives normally represent a more abstract, coarser description of the landscape defined by a function: a description that encompasses a larger swath of space and/or time. The first derivative, or slope, describes the incline in the immediate vicinity of a point. The second derivative, or curvature, summarizes the relationship between several inclines, and the third derivative, often called *texture*, denotes the relationship between several curves. For example, a surface with *bumpy* texture consists of many areas of abruptly changing curvature, and the concept of bumpiness normally refers to a region: it is typically not the property of a single or small number of neighboring points.

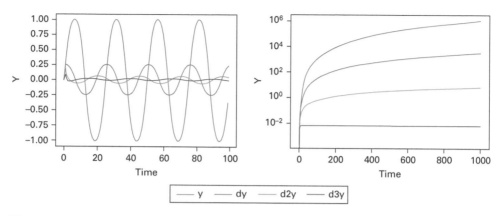

Figure 3.10
Outputs of neuron Y and the three derivative neurons when the delayed-inhibition network (figure 3.8, bottom) processes two different input sequences: $y = sin\left(\frac{t}{4}\right)$ for t = 0, 100 (left), and $y = \left(\frac{t}{10}\right)^3$ for t = 0, 1000 (right).

The plots of figure 3.10 show that reasonably accurate derivatives result from a chain of delayed-inhibition units, but the reverse process should also be possible: the higher levels should be able to drive the lower levels as long as each level receives and integrates inputs from its immediate high-level neighbor. Thus, the highest level should, in essence, predict the values of all lower levels. Additionally, the higher levels should be able to do this while updating at slower timescales. Figure 3.11 illustrates this effect: when the third derivatives corresponding to those from figure 3.10 are fed into the top neuron of figure 3.11 (left) and then propagated downward to neurons (that retain a fraction of their previous value between timesteps), the outputs of each neuron closely mirror those plotted during the inverse process of differentiation from Y upward (in figure 3.10).

When differentiators and integrators are combined into a layered architecture, the integrator at one level can predict the value at its lower neighbor. Figure 3.12 (left) shows a hierarchical composition of these differentiator-integrator pairs, with their key interface being the error unit at each level.

Moving to the right side of figure 3.12, after a brief spin-up period during which the upper levels integrate signals from the lower levels, the network's stable state displays predictive inputs from above that cancel the state values below, yielding small errors. This happens despite the fact that upper levels update much slower than lower levels. In essence, the upper levels project stable long-term expectations (e.g., averages) that give (imperfect but) reasonable estimates of lower-level states, which display more dramatic flux than their above neighbors.

This extremely simple model is intended as no more than a basic abstraction of what a hierarchical brain can do: integrate upward-flowing derivative signals at one level and then use those aggregates as coarse predictions of future states below, while simultaneously providing derivatives of one's own state to even higher levels, running at still slower timescales. The net result is a predictive hierarchy, a more detailed biological version of which appears in the neocortex, as described in chapters 5 and 6.

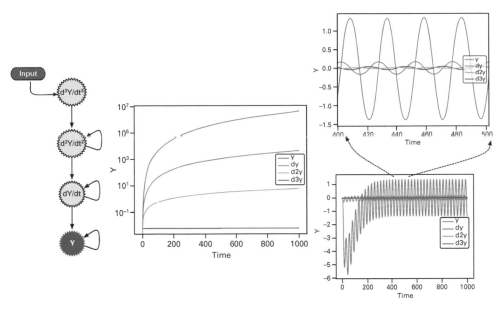

Figure 3.11
(Left) A simple integrator network where weighted output of each upstream neuron combines with a fraction of its downstream neighbor's previous value. Time series (t = 0 to 1000) of each neuron's value when continuously supplied with the following inputs at the top, third-derivative neuron: (Middle) $\frac{\partial^3 y}{\partial t^3} = \frac{6}{1000}$, and (Right) $\frac{\partial^3 y}{\partial t^3} = -\left(\frac{1}{4}\right)^3 cos\left(\frac{t}{4}\right)$. In both runs, the ratios of time constants from Y upward are 1:2:3:4.

3.3 Birth of a PID Controller

As discussed earlier, the combination of derivatives and integrals (sums, averages) directly supports both prediction and control, as encapsulated in the classic equation for a PID controller (repeated from chapter 2):

$$u_t = k_p e_t + k_d \frac{\Delta e_t}{\Delta t} + k_i \sum_{j=0}^{t} e_j \qquad (3.1)$$

In short, the error term (difference between goal and current, or predicted, state) undergoes three basic operations (scaling, differentiation, and integration) whose results then combine to produce the control output, which, as detailed earlier, corresponds to a prediction of a future state or error. A simple neural network to perform this calculation appears in figure 3.13.

Nervous systems are replete with all of the primitives that constitute PID and other controllers. To compute error, at least as a first approximation, a network merely needs a comparator that subtracts one input (e.g., the prediction) from another (e.g., the target). Hence, if predictive synapses inhibit the comparator while target synapses excite it, the combination embodies a comparison that yields predictive error. To integrate signals over a longer time frame than that of a single depolarizing or spiking event, an individual neuron merely needs physicochemical properties that dictate a larger time constant (than that of a spike). These *slower* neurons abound in all nervous systems. Another common neural

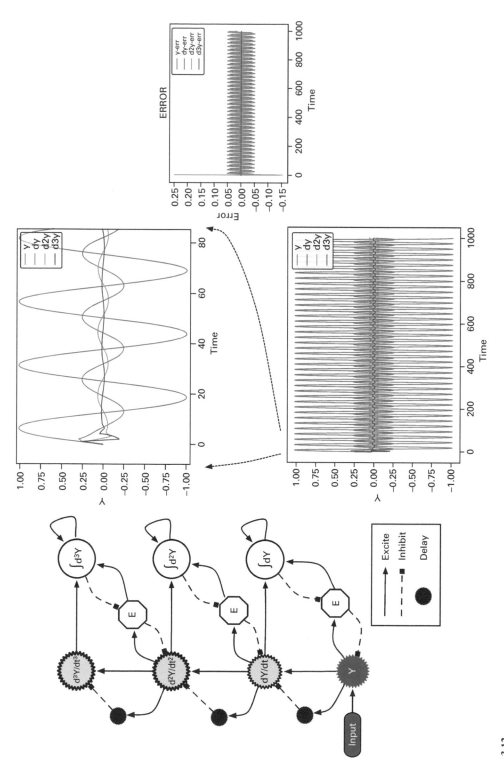

Figure 3.12

(Left) A simple network that combines delayed-signal derivatives and integrators (of derivatives) to perform predictive coding; the E nodes compute prediction error. (Right) Results of the network when fed a time series of values: $y = sin\left(\frac{t}{4}\right)$ for $t = 0$, 1000. The ratios of time constants from Y upward are 1:4:8:16, indicating that the upper-layer integrator $\int \frac{\partial^3 y}{\partial t^3}$ updates at $\frac{1}{16}$ the speed at which the bottom layer receives inputs. The upper plot (of Y and its derivatives) is a magnified portion of the first 80 timesteps of the lower plot, while the rightmost plot is of the error units over the full 1,000 timesteps.

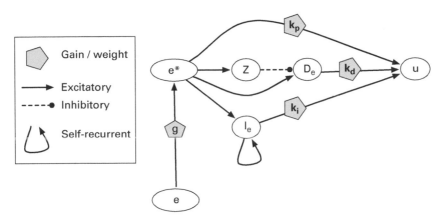

Figure 3.13
A neural network implementation of a PID controller. The original error (e) is scaled by a gain/weight (g) during transfer to gateway error neuron, e*. Each of the constants from equation 3.1 appear as weights on connections to the output neuron, u. The self-recurrent loop of neuron I_e performs basic integration, since I_e's output from timestep t becomes part of the input at step t+1. The delayed inhibition in the e*-Z-D_e chain computes a derivative.

mechanism for caching activation histories is recurrence, wherein neuron A feeds into neuron B, which then feeds back to A. The time lag involved in this double transfer ensures that A receives information colored by its past activity, which it combines with signals from other sources that are more strongly representative of the present. As for the constant terms (k_*) in figure 3.13, these are captured by the total number, location (proximal or distal), and strength of synapses linking one unit, such as a comparator, to another, for example, an integrator.

Perhaps most interesting are the delay lines that facilitate gradient calculations. As discussed below, brains are a mixture of excitatory and inhibitory neurons (at an overall ratio of about 5:1, with excitation in the majority (Kandel, Schwartz, and Jessell 2000)), and most local neural circuits also contain an assortment of both types. When axons from excitors of area A grow into area B, they may target B's excitors, inhibitors, or both. And the inhibitors tend to synapse on the excitors (along with other inhibitors). Collectively, these connections support a circuit similar to that involving e*, Z, and D_e in figure 3.13: the direct connections from A to B's excitors mirrors the e*-D_e link, while the inhibitors of B resemble Z in that they both delay and invert the signal from A's to B's excitors. The inhibitors delay transmission simply by being one extra link in the synaptic chain from A's excitors to B's. Of course, stringing several such circuits together (A to B to C) can realize higher-order derivatives as well.

3.3.1 Adaptive Control in the Cerebellum

One area of the brain often characterized as a controller is the cerebellum, which has a well-established role in the learning and control of complex motions (Kandel, Schwartz, and Jessell 2000; Bear, Conners, and Paradiso 2001), as well as cognitive skills such as attention and speech production. In general, it seems to have a lot to do with timing aspects of various skills; and because of the inherent delays in nervous systems, proper timing requires prediction.

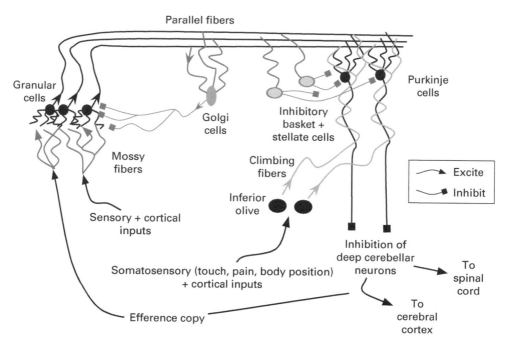

Figure 3.14
Basic anatomy of the mammalian cerebellum, based on images and diagrams in Kandel, Schwartz, and Jessell (2000); Bear, Conners, and Paradiso (2001); and Rolls and Treves (1998).

As shown in figure 3.14, the cerebellar input layer, consisting of granular cells, receives a variety of peripheral sensory and cortical signals via mossy fibers stemming from the spinal cord and brainstem, and realizing a mixture of delays. As the most abundant neuron type in the mammalian brain—the human cerebellum contains approximately 10^{11} (Kandel, Schwartz, and Jessell 2000)—these granular cells appear to manifest expansion coding of sensory and *corollary-discharge* signals, that is, efference copies of motor commands. The tendency of granulars to laterally inhibit one another (via nearby golgi cells) characterizes them as sparse-coding context detectors (Rolls and Treves 1998). Since delay times vary along the mossy fibers, each context has both temporal and spatial extent.

One parallel fiber (PF) emanates from each granular cell and synapses onto the dendrites of many Purkinje cells (PCs), each of which may receive input from 10^5 to 10^6 parallel fibers (Kandel, Schwartz, and Jessell 2000). The PFs also synapse onto basket and stellate cells, both of which inhibit nearby PCs. Hence, the granulars can have both a direct excitatory effect on Purkinje cells and a delayed inhibitory influence. As described above, this allows PCs to detect temporal gradients of the contexts represented by the granulars.

Since the Purkinje outputs are the cerebellum's ultimate contribution to motor and cognitive control, the plethora of granular inputs to each Purkinje cell appears to represent a complex set of preconditions for the generation of any such output. These antecedent-consequent rules are adaptable, since the PF-PC synapses yield to both long-term potentiation (LTP) and (more prominently) long-term depression (LTD) (Kandel, Schwartz, and Jessell 2000; Rolls and Treves 1998; Porrill and Dean 2016). Adaptation, particularly LTD, of the PF-PC synapses is triggered by inputs from climbing fibers of the inferior olive, which transfer information such as pain signals from the muscles and joints directly influenced by those fibers' corresponding PCs. Climbing fibers exhibit a simple version of supervised learning (Doya 1999) wherein the olivary pain signal modulates the PC's output (which appears to manifest a combination of tracking and prediction error) to drive LTD of the PF-PC synapses. In this way, any combination of predicted-state, goal-state, and sensory information that produces a particular action (that leads to an immediately undesirable outcome) will have a reduced ability to incite that action in the future.

Although learning via the PF-PC synapses involves both LTP and LTD (Porrill and Dean 2016), the latter has garnered the most attention in neuroscience research. The prominence of LTD may stem from the sheer density of afferent parallel-fiber synapses onto each Purkinje cell, and thus the need to adaptively neutralize as many as possible to form viable context-action rules. This hints of Edelman's (1987) *neural selectionism* (discussed in chapters 6 and 7), in which many synapses form early in life but are then gradually pruned by the LTD induced by the agent's experiences.

It is also worth noting that outputs from PCs have inhibitory effects on their postsynaptic neighbors. Thus, the arduous process of tuning brains to achieve motor and cognitive skills involves figuring out *what to turn off* in a given context. This paints the skeletomuscular system as a collection of overeager motor units that are gradually tamed and coordinated by the cerebellum. Somewhat counterintuitively, the simplest behaviors often require the most complex neural activity patterns. For example, it takes a much more intricate combination of excitatory and (particularly) inhibitory signals to wiggle a single finger (or toe) than to move all five. Hence, the tuning of PC cells to achieve the appropriate inhibitory mix is a critical factor in basic skill learning.

The earliest cybernetic models of the cerebellum date back to the work of David Marr (1969) and James Albus (1971). Known as the *Marr-Albus model*, it has many interpretations, including that of an adaptive controller (Fujita 1982), which employs a control element known as an adaptive filter (aka forward model) to predict future states given current states and actions. As pointed out by Porrill and Dean (2016) in their investigation into the Marr-Albus and other models of cerebellar control, contemporary neuroscientific evidence still gives modelers considerable interpretive freedom. The following brief description exercises a bit of that flexibility while illustrating the use of prediction and control in cerebellar function.

Figure 3.15 portrays a hybrid anatomical-functional model of the cerebellum, with abstractions of the basic topology framing the functional modules. Be aware that any attempts to map controller components to the anatomy of any brain region involve a great degree of speculation.

Starting with the mossy fibers and granular cells, their inputs include many aspects of context: sensory (including proprioceptive) signals, desired / goal states of the brain-body-environment coupling, and corollary discharges (aka efference copies) from motor and premotor regions that indicate the impending action. Since this region appears to integrate information across time, via both the differential delay lines on the mossy fibers and the recurrent loop from granulars to parallel fibers to golgi cells and back to granulars, it seems well-equipped to make predictions of future states based on a recent time window of sensory input plus the efferent motor copy. Thus, this area constitutes a forward model that can be classified as an adaptive filter if several of its parameters admit tuning. And, indeed,

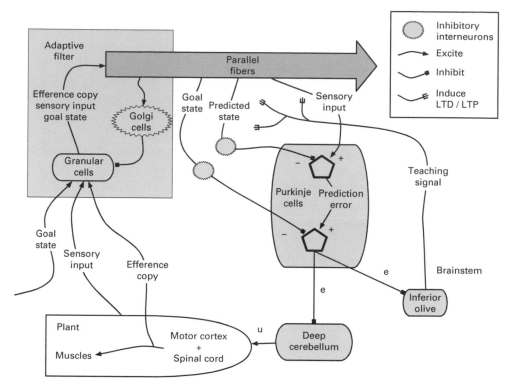

Figure 3.15
Model of cerebellum as an adaptive controller. Controller modules are shaded, while the animal's body, which constitutes the *plant* of cybernetics, is unshaded. Locations of the comparators, adaptive filter, control error (e), and control output (u), are estimates; these may be more widely distributed throughout the cerebellum and brain as a whole.

many synapses in the cerebellum admit LTP and LTD. Since the main evidence of cerebellar synaptic change comes from the PF-PC link, these too might be included in the adaptive forward model. At any rate, by the time signals reach the soma of Purkinje cells, they should encode a predicted state along with the current state and goals.

The Purkinje cells then act as comparators. Since PCs do not feed into one another in series, the vertical string of comparisons shown in figure 3.15 would need to invoke the rich web of PC dendrites, which many researchers propose as the source of complex computations well beyond that of simply transferring all afferent signals to the soma for summation (Hawkins and Ahmad 2016). Taken in succession, for ease of explanation, the first comparison involves the most recent sensory information (constituting a current state) and a prediction of that state (based on an earlier state and the efference motor copy).

For many tasks, such as maintaining stable views of the environment despite head movement, the prediction (e.g., of image movement) derived from the impending action (e.g., head rotation), must be subtracted from the raw image (of drastically shifting surroundings) to yield the actual perception (of a stationary world). Thus, subtraction of a prediction (including the efference copy) from the raw sensory reality produces an accurate picture of the current situation. The Purkinje area's second comparison involves the prediction-adjusted state and the goal, giving the ultimate output error. This is the standard error term in a feedback controller, the one used to determine the next action. In figure 3.15, this error information emanates to deeper cerebellar areas, where it is converted into a control signal (u) for directing motor activity. Sensory signals from the body along with the efference copy from the motor system then complete the feedback loop via the mossy fibers.

Evidence of the Purkinje cells' role as comparator of predictions, goals, and efference copies to sensory reality extend back in evolutionary history to a diverse collection of primitive (extant) fishes, all of which exhibit parallel fibers and Purkinje cells in cerebellar-like structures. As described by Bell and colleagues (1997), the inhibitory stellate neurons near the PF-PC interface can invert prediction signals such that the net Purkinje output embodies a comparison of reality to expectations. In addition, these proto-cerebellums display climbing fibers that initiate anti-Hebbian modifications to the PF-PC synapses.

3.4 Detectors and Generators

Up to this point, we have viewed prediction in terms of scalar values, their derivatives, and their integrals. This may apply to some simple nervous systems, where the outputs of isolated neurons have significant semantic value, but most advanced neural systems employ population codes: the activity levels of hundreds or thousands of neurons may constitute a salient signal or representation. Patterns in one network region (e.g., layer) can then cause patterns in another, with designations of these regions as higher or lower than one another (though sometimes arbitrary) framing the activity as bottom-up or top-down.

For example, figure 3.16 (left) displays a bottom-up scenario in which a pattern appears at a lower level and then promotes firing of a detector neuron (Z) due to strong synaptic connections between the three constituents and Z. Conversely, Z can function predictively if its activity causes neurons A, D, and E to activate, as shown on the right of the figure. The generative direction also characterizes motricity, wherein higher-level activities lead to the firing of motor neurons and the contraction of particular combinations of muscles.

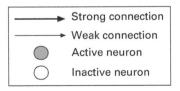

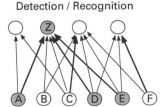 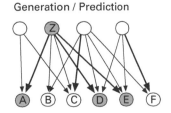

Figure 3.16
Two directions of signal flow in a neural system. (Left) Bottom-up transmission from, for example, sensory levels to higher regions. Neuron Z detects pattern A-D-E. (Right) Top-down signaling from higher to lower levels. Neuron Z predicts pattern A-D-E.

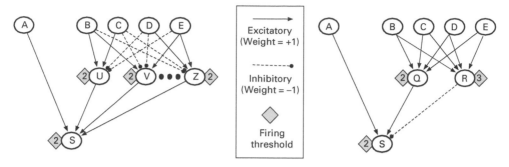

Figure 3.17
Simple neural networks for detecting the concept *A and exactly two of the other four inputs*, where inputs A–E have binary values (0 or 1) and feed into nonlinear neurons that output a 1 if and only if their sum of weighted inputs is equal to or above the associated firing threshold. (Left) A complex network solution that checks each of the six mutually exclusive pairs of B–E. (Right) A simpler network that requires Q to fire (indicating two or more of B–E) and R to remain silent (indicating that less than three of B–E have fired).

As the patterns to detect or predict become more complex, so too do the neural circuits for handling them. One key to complexity is a mixture of excitation and inhibition: too much of one or the other leads to all-or-nothing activity across a neural region, a situation that typically conveys little useful information. Although a single node in an artificial neural network may stimulate some neurons and inhibit others, these two activities normally require different types of neurons in brains. Thus, an excitatory neuron can inhibit another neuron only by first stimulating an inhibitory interneuron. The sample networks described below follow the *artificial convention* to simplify the diagrams, but be aware that inhibitory connections would actually require additional intervening neurons in a biological network.

Multiple layers of neurons provide another complexity enhancement. Consider the circuits of figure 3.17, where unit S acts as the detector for a particular activity pattern among the five sensory units. This computation requires excitatory and inhibitory interneurons, although their cardinality and connection topology can vary. On the left of the figure, the interneurons U–Z each detect one of the six possible combinations of B–E. Notably, this demands many inhibitory connections to ensure that exactly two of the four sensors have fired. However, a simpler circuit (on the right), with only one inhibitory link, also solves the problem. This requires neuron Q to detect *two or more* and neuron R to detect *three or more* such that neuron S fires when A and Q fire, but R does not. Imagine the savings in both interneurons and inhibitory connections, by using a slightly modified version of the rightmost circuit, if the problem were expanded to *A and any three of the other fifty*. A few logical tricks can save a lot of resources when building detectors.

The situation gets more complicated for generators. The circuits in figure 3.18 are designed to activate A and exactly two of B–E whenever unit S fires. Think of this as playing a chord on an instrument such as a clarinet, where A controls the thumb, which needs to remain active to balance the instrument, while two of the four fingers need to change position to open or close particular holes. The difficulty stems from the inversion of the detector problem, which mapped many sensory possibilities to one detector. Now, the mapping goes from one *intention to act* to many motor alternatives, only one of which can activate. This type of circuit typically requires a lot of inhibition along with more precise timing considerations.[1]

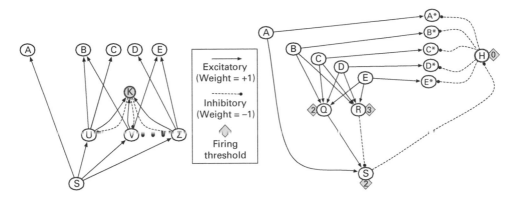

Figure 3.18
The inverse situation of figure 3.17: Each cell A–E represents a motor unit, and when S activates, it should trigger a complex activity in which unit A activates along with exactly two of the B–E units. (Left) Each of the U–Z units codes for one unique pair of the B–E units. The signal from S presumably activates U–Z asynchronously such that exactly one of them (e.g., U) fires first. That triggers the B–C pair while also exciting neuron K, which immediately inhibits all of the U–Z neurons. Thus, timing and widespread inhibition become crucial. (Right) Units A–E now represent premotor neurons that stimulate the corresponding motor units A*–E* (each of which has a firing threshold of 1). As in figure 3.17, unit S functions as a recognizer of the *A and exactly two of B–E* pattern. Premotor units A–E fire randomly, with no motor effects until S fires, thus *unlocking*, via disinhibition across H, the motor units, which require input from their corresponding premotor unit to fire. Inhibitor neuron H requires no input to remain active. Note that this scales linearly to any number of premotor and motor units, with each new pair requiring a small constant number (4) of additional connections.

One approach (figure 3.18, left) involves interneurons U–Z, which again represent each of the mutually exclusive alternatives; but proper behavior now depends on asynchronous firing and fast inhibition of the later-activating alternatives, as explained in the figure caption. This solution scales very poorly (as does the leftmost detector circuit in figure 3.17) for networks with many motor units. It seems unlikely that any one-to-many circuit could perform this operation without many interneurons and considerable inhibition.

Fortunately, the problem can be reformulated as one of detection and disinhibition, as shown on the right of figure 3.18. Now units A–E function as randomly firing premotor neurons, whose *intended* activation pattern runs through units Q, R, and S, with S once again acting as a detector that now *gates* the premotor intention forward to the motor units, A*–E*, in much the same way the basal ganglia appears to gate premotor intentions through disinhibition (Houk, Davis, and Beiser 1995). In this case, the premotor signals are blocked from producing motor activity until the gating condition is satisfied. Thus, instead of working from the default assumption in connectionism that all units are inactive until stimulated by input or an upstream neighbor—an assumption that works well for pattern detectors—a network designed to produce complex actions benefits from assuming the random percolation of intentional units whose downstream signals are normally blocked but occasionally released (disinhibited).

This approach scales well, as illustrated by figure 3.19, where one network can realize different intricate actions, with each requiring only a different pattern detector, not a combinatorial explosion of interneurons and inhibitors. However, this design has at least one major flaw: temporal overlap between valid and invalid patterns. For example, assume that the premotor pattern A-B-C has just fired. This triggers unit S and thus disinhibits *all* of the motor units, but only three of them, A*, B*, and C*, have enough positive stimulation to

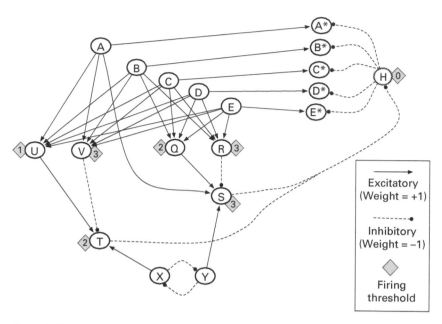

Figure 3.19
A network for evoking one of two mutually exclusive patterns using motor units A*–E*: (1) Activate exactly one or two of the motor units (when unit X fires), and (2) activate unit A* and exactly two of units B*–E* (when unit Y fires). This assumes random firing of the premotor units, A–E, such that when X or Y fires, the detector T or S, respectively, has enough base stimulation to fire if and only if its other afferents (U and V, or Q and R, respectively) have the appropriate settings. Firing of S or T then deactivates the inhibitor unit, H, thus releasing all motor units (which have a firing threshold of 1) from inhibition and allowing any stimulated by their premotor counterpart to fire.

actually fire. So far so good. But what if unit D spontaneously fires just after A-B-C and S? Since disinhibition spans all motor units, D* would also fire, producing an improper motor activity.

To remedy these types of problems, brains have evolved a wide range of oscillatory patterns, from 0.01 to 600 Hz (Buzsaki 2006). Clusters of neurons firing (more or less) synchronously, but without external influence, constitute internal rhythm generators whose periodic signaling governs the firing behavior of their efferents. In figure 3.20, the Z cluster represents an oscillator whose periodic inputs to premotor units A–E is just enough to push them to their firing thresholds in situations where they have also been stimulated by one other pathway (or by spontaneous depolarization). Hence, for units A, B, and C to fire unit S, they must first reach their firing thresholds with the help of Z's input, which only occurs periodically, at the top of each cycle. When this happens, all motor units briefly disinhibit and A*–C* can fire. However, when unit D exhibits spontaneous activity shortly afterward, the Z cycle has already declined, and D cannot fully depolarize and therefore cannot stimulate motor unit D*.

Essentially, the oscillator *sweeps up* all units that have been partially active during its ascending cycle by infusing them with enough extra signal to push them over their thresholds, thus gating a pattern forward (to the detectors and motor units) via a synchronous round of premotor firing. But any premotor units that begin depolarizing after the peak (e.g., unit D in the figure) will either die out, due to the short latency of their spontaneous activity, or have to wait until the next peak (if they can maintain their semi-depolarized state).

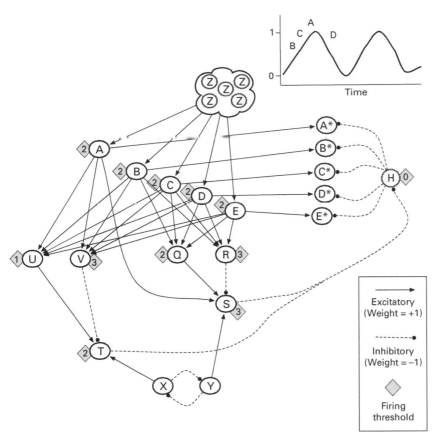

Figure 3.20
A multiple-action network in which premotor units A–E require input from the oscillatory neuron cluster (Z's) along with their own spontaneous activity to reach their firing threshold of 2. The graph in the upper right tracks the synchronous activity of the oscillating cluster, which fires (sending a 1 to the premotor units) only at the peak of each cycle. The letters A–D on the graph indicate the time points at which premotor units A–D are spontaneously active but not spiking (i.e., depolarizing but not yet firing due to a higher threshold than that exhibited by the same neurons in figure 3.19.)

Beyond this simple example, in real brains, oscillators of varying frequencies can sweep up signals exhibiting more or less temporal disparity. A low-frequency rhythm can bundle semi-active neurons over a broad temporal window (assuming that each has a slow time constant), whereas quick oscillators impose stricter time constraints on potential cell ensembles. Thus, given the basic relationships between time and space, a low-frequency cycle can help coordinate the firing of neurons that interact over longer distances but would have a hard time synchronizing otherwise. In short, the brain's cycles both enable and enforce synchronicity such that the neural network can operate in a more discrete fashion, gating in well-established activity patterns with a minimum of noise.

When viewing the brain as a predictive machine, the top-down generative phase of network behavior takes center stage. Since most high-level concepts can be realized by a combinatorial explosion of lower-level patterns, the circuitry for prediction in living organisms probably demands considerable inhibition, disinhibition, and oscillation to ensure that only one of these options gets chosen at any one time, without interference from other patterns. Thus, at first glance, it appears that the computational demands of generation / prediction exceed those of detection / recognition. However, in theory, the brain must also confront the fact that an infinite number of interpretations exist for any sensory state, so sorting them out may also require its fair share of resources.

3.4.1   The Hippocampus

In mammals, the cerebral hemispheres comprise the neocortex, at the top, and three main subcortical structures: amygdala, basal ganglia, and hippocampus (Kandel, Schwartz, and Jessell 2000). The neocortex is a thin sheet of six-layered cortical columns, which chapter 5 explores extensively as a sophisticated, modular architecture for predictive coding.

Although reptiles, amphibians, and birds lack a neocortex, they have cranial regions analogous to these three subcortical areas (Striedter 2005). In particular, they all have a hippocampus (HC), which is essential for memory formation (Squire and Zola 1996; Andersen et al. 2007). The neocortex is often viewed as the *crowning* (literally) achievement of mammalian brain evolution, the cerebral matter that (somewhat) justifies the moniker *higher intelligences* bestowed on humans, monkeys, and even rats. Although the hippocampus resides below the neocortex, and is more primitive evolutionarily, it seems to occupy the top layer of the *functional* cortical hierarchy; and it exhibits a very sophisticated form of predictive coding.

Comparative brain anatomy (Striedter 2005) clearly reveals that inputs to HC come less from low-level brain regions and more from higher regions as one ascends from amphibians to reptiles to birds and finally to mammals (Striedter 2005), where the entorhinal cortex (EC) serves as an exclusive gateway to HC: almost all signals going into and out of HC go through EC (Rolls and Treves 1998; Andersen et al. 2007). Neural firing patterns in EC have already been through many levels of processing, so they represent reasonably abstract, multimodal (i.e., involving combinations of visual, auditory, haptic, olfactory, and so on) information. When fed into HC from EC, these signals distribute to several different areas, but with different effects.

As shown in figure 3.21, axons from EC layer 2 connect directly into the dentate gyrus (DG), which feeds into CA3 and then on into CA1. As detailed by Rolls and Treves (1998), DG performs pattern sparsification via abundant lateral inhibition: neurons compete with one another to fire, with the active neurons loosely representing principal components. These sparse patterns then enter CA3 via (relatively) direct lines from DG to CA3 pyramidal cells: the fanout is approximately only 1:12 (Andersen et al. 2007), which is quite low for interpopulation connections. CA3 has very high recurrence—the most of any brain region; each CA3 pyramidal connects with about 5 percent of the others via excitatory synapses (Rolls and Treves 1998; Kandel, Schwartz, and Jessell 2000). Thus, CA3 seems to act as a pattern-completing area; and when those patterns have significant temporal scope (as most patterns at higher, slower levels do), pattern completion manifests prediction. In this way, CA3 appears to function as a fill-in-the-blank station for episodic memories (i.e., those of sequences of experiences) cued by a few key hints from DG.

For predictive coding, the pivotal area is CA1, where direct inputs from EC (layer 3) meet the sequence predictions from CA3 (that originated in layer 2 of EC). As shown in figure 3.22, these signal lines meet the dendrites of CA1 pyramidal cells at different locations. The direct lines from EC synapse distally, in a layer containing very little

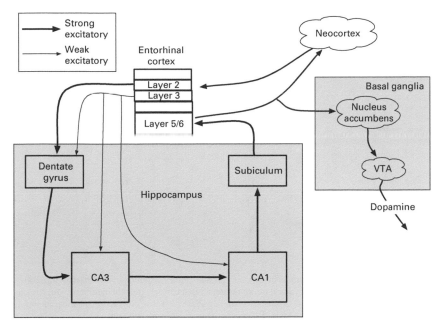

Figure 3.21
Basic topology of the hippocampus (HC), its gateway region (the entorhinal cortex, EC), and other primary sources and targets of HC signals.

inhibition. Hence, these *reality* signals excite the pyramidals. Conversely, the CA3 inputs arrive proximally, near the somas of CA1 pyramidals, in an area very dense with inhibitory interneurons. Thus, although the link between CA3 and CA1 along the Schaffer collaterals is strongly excitatory (Kandel, Schwartz, and Jessell 2000), the net effect may ultimately be inhibitory due to the ensuing interneuron activity. This has led several researchers to propose CA1 as the comparator region for predictions (from CA3) and reality (from EC) (Ouden, Kok, and Lange 2012; Lisman and Grace 2005), with the resulting prediction error representing the level of *novelty* in the current input signal from EC to HC.

As shown in figures 3.21 and 3.23, output from CA1 eventually returns to layers 5/6 of EC on its way back to the neocortex. The novelty signal thus contributes to an abstract prediction of its own (of an expected sequence of upcoming experiences), in much the same way that upward-traveling prediction errors in cortical columns pass through transformative connections before returning downward as feedback predictions. Perhaps more importantly, the novelty signal also reaches the ventral tegmental area (VTA, upper right corner of figure 3.21), which broadcasts dopamine when sufficiently stimulated, and this neuromodulator then stimulates learning.

In summary, those prediction errors that resist being *explained away* by feedback predictions will work their way up the cortical hierarchy to EC, as proposed by Hawkins (2004). The hippocampus then gets one last chance to reconcile the information with its archive of abstract sequences (recalled via activity in DG and CA3). Any remaining error constitutes novelty, which the brain will then try to learn: the neural assemblies that were recently active in the neocortex will be encouraged, by dopamine from VTA, to bond. This learning process continues during sleep, when interaction between HC and neocortex remains strong.

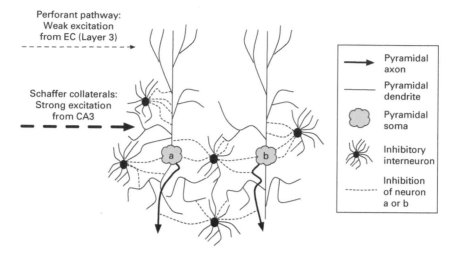

Figure 3.22
General cytology of area CA1 of the hippocampus based on diagrams and descriptions in Andersen et al. (2007). Pyramidal cells in CA1 receive weak excitatory inputs from the entorhinal cortex via distal dendrites, which reside in a layer with very few inhibitory interneurons. Conversely, strong excitation from CA3 via the Schaffer collaterals enters much closer to the soma (a and b), in layers with a much higher density of interneurons of many types, but all inhibitory.

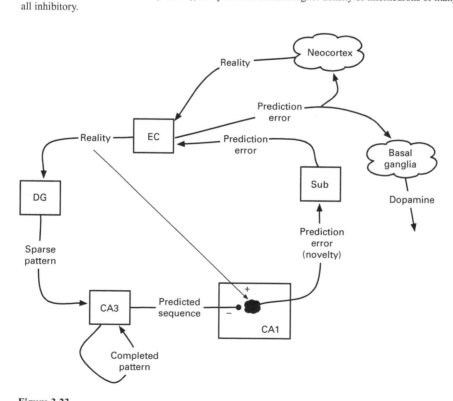

Figure 3.23
Functional diagram of the hippocampal formation as a model of predictive coding. EC = Entorhinal Cortex, DG = Dentate Gyrus, Sub = Subiculum. As hypothesized by several neuroscientists (Ouden, Kok, and Lange 2012; Lisman and Grace 2005), area CA1 acts as a comparator of a prediction from CA3 and reality signal from EC. This relies on the assumption that the excitatory Schaffer collateral pathway from CA3 to CA1 synapses on many of the inhibitory interneurons near the soma and proximal dendrites of CA1 pyramidal cells.

As a domestic example, when my doorbell rings, I immediately begin to conjure up predictions as to the imminent visit. The current context, for example, Halloween night, combines with the doorbell information in driving CA3 to dredge up a memory of trick-or-treaters, which constitutes my expectation (and may cause me to rummage through the house for candy and a makeshift costume). However, when I open the door to find my son, who left his house key at the gym, the deviation of that reality (sent directly to CA1 from EC) to the CA3-generated expectation startles me and may lead to changes of my Halloween memories, twenty years later, I will still tease my son about the time he forgot his key on beggar's night and was met at the door by Socrates and a handful of miniature candy bars.

3.4.2   Conceptual Embedding in the Hippocampus

In 2005, the research group led by May Britt and Edvard Moser discovered *grid cells* in an area of the hippocampal gateway known as the medial entorhinal cortex (MEC) (Hafting et al. 2005). The discovery earned them the 2014 Nobel Prize in medicine, an honor that they shared with their mentor, John O'Keefe, who discovered *place cells* in the hippocampus three decades earlier (O'Keefe and Dostrovsky 1971). The differences between grid and place cells, along with their interactions, provide an interesting backdrop for navigation, prediction, and intelligence in general.

Place cells are neurons in hippocampal regions such as CA3 and CA1 that tend to fire only when the organism resides in (or approaches) a particular location (Andersen et al. 2007). Hence, each place cell constitutes a detector for some fairly localized spot in the environment. In contrast, grid cells have the fascinating property of detecting all spots at the corners (and center) of hexagonal patterns laid out across an environment. Thus, a particular grid cell systematically cycles through active and inactive phases as the animal (to date, typically a rodent) moves about its environment. As depicted in figure 3.24, if an experimenter marks the spots where that cell fires on a map of the environment, a pattern of equilateral triangles arises; and together, these form hexagons: the receptive fields of grid cells are hexagonal grids.

Figure 3.24 also illustrates that as the animal moves in a fixed direction, the pattern of grid-cell activity should essentially follow a cyclic pattern as the corners of overlapping triangles are visited. As explained more thoroughly by the Mosers and colleagues (2014), inputs to grid cells from both head-direction and velocity cells are believed to push activation patterns around MEC, producing cyclic activity bumps among grid cells with similar receptive fields. The faster the animal moves, the more rapidly does the bump. Interestingly, the Moser Lab has also found evidence of a special cluster of MEC cells that represent context-free velocity (Kropff et al. 2015), and they too appear to be predictive: their activity correlates better with future than with past or present speed.

The precise purpose(s) of grid cells and their hexagonal receptive fields have yet to be proven, but most believe that they provide a form of mental global positioning system (GPS) when combined with place cells. Basically, when an animal resides in a given spot, S, and moves in a known direction with a known speed, then it can estimate its new location (S*) without receiving much sensory feedback, for example, in the dark. In effect, the animal makes a *prediction* via a process of dead reckoning wherein grid cells support sensory-feedback-free transitions between place cells.

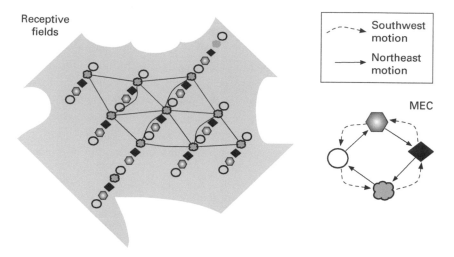

Figure 3.24
The basic behavior of grid cells. (Left) An environment (shaded background) labeled with small shapes denoting the grid cell that fires when the organism resides in that particular location. The hexagonal pattern depicts a cluster of places (i.e., the corner points of each triangle) where the neuron, represented by a shaded, flower-shaped pattern, will fire, aka that neuron's *receptive field*. Similar hexagonal receptive fields exist for each of the four neurons. (Right) Relationships between four grid cells for achieving the receptive fields on the left. When the agent moves in a northeast (southwest) direction, the solid (dashed) arrows portray the sequence of grid-cell stimulation: the activation of any grid cell combines with sensorimotor inputs to stimulate its neighbor. When neighboring grid cells have similar receptive fields, the cyclic activation pattern is that of a moving bump or blob in a multidimensional neural space.

Underlying this dead reckoning is the ability to triangulate a location using multiple grid cells. MEC houses grid cells whose receptive fields vary in spatial frequency, phase, and rotation (Moser et al. 2014). Thus, as shown in figure 3.25, the receptive fields of different cells may intersect at only a few points, and the more grid cells involved, the fewer points of mutual activation.

The information encoded in the firing of a single grid cell is rather diffuse, indicating only that the animal resides near one of the many triangle corners in its hexagonal grid. But when two or more grid cells (of different frequency, phase, and/or rotation) fire, this helps pinpoint the actual location. In the cartoon example of figure 3.25, it takes only two grid cells to reduce the number of possible locations to three, and complete disambiguation results from the addition of a third grid cell. In reality, this reduction of possibilities may require four or more grid cells, but the basic process seems quite plausible.

Thus, combinations of active grid cells may indicate unique locations, that is, *places*, and connections from MEC to the interior of the hippocampus (areas CA3 and CA1) may constitute detector circuits for those places. As shown in figure 3.26, both place cells, P_a and P_b, have incoming connections from a set of three different grid cells. When one of these grid triples fires, so too does the corresponding place cell. The figure omits many details, including the recurrent pathways that connect CA3 and CA1 back to MEC (via the subiculum) (Moser et al. 2014; Andersen et al. 2007). This feedback could enable an active place cell to excite a group of related grid cells: those that triangulate that place cell.

The animal then makes predictions of places, either those in the future or in the present but invisible due to limited or delayed sensory input, by running these circuits in both directions.

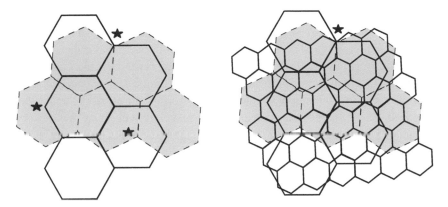

Figure 3.25
Triangulation of environmental locations using grid cells. (Left) Each set of hexagons (one with thick borders and the other with dashed borders and gray interior) represents the receptive field of a different grid cell. Starred points are those where the fields intersect, either at the corners or centers (not drawn) of hexagons. (Right) Addition of a third grid field (in this case of higher frequency, i.e., smaller hexagons) reduces the number of three-point intersections to the single starred location.

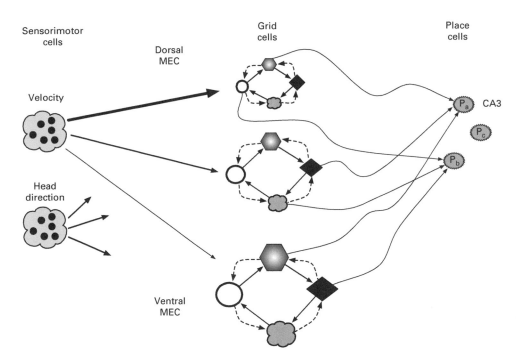

Figure 3.26
Interactions among sensorimotor information, grid cells, and place cells. In moving from the dorsal to ventral MEC, grid cells have lower-frequency (higher amplitude) receptive fields denoted by larger cell icons (Moser et al. 2014; Andersen et al. 2007). For any given frequency and orientation, the grid cells representing different spatial phases (offsets) might interact via cyclic excitation (as denoted by the three bidirectional cycles). As a simplified interpretation, the active head-direction cells could *select* the appropriate cycle at each frequency level, while velocity inputs would drive the activity bump around those focal cycles. The thickness of arrows emanating from the velocity group indicate strength of influence (e.g., synaptic strengths) such that a given velocity should drive a high-frequency grid cycle harder (faster) than a low-frequency loop.

Initially, the agent surmises its current location via sensory cues, such as spatial landmarks, olfactory or auditory signals, and so on. Detectors for these sensory data also serve as inputs to the hippocampus, indirectly via MEC and other regions of the entorhinal cortex. Hence, these detectors can activate place cells (P*), which, in turn, can *reset* the grid system such that the triangulators (G*) of all active place cells also begin firing.

Next, the animal moves but without receiving sufficient sensory information, either due to sensory deprivation (e.g., darkness) or owing to the relatively long delays in sensory processing compared to those of action. However, the proprioceptive input enables the agent to estimate its velocity and egocentric head direction, and this information, in turn, can push the activity bumps associated with G* through several steps of their associated cycles. The final locations of those bumps then represent a set of active grid cells that map to a new set of place cells that represent the predicted location. The later arrival of external sensory information may then confirm or refute that prediction, but even if incorrect, the prediction may provide advantages over the completely naive state.

A major question in cognitive neuroscience is whether or not the grid-place-cell network could be employed for tasks other than navigation (Bellmund et al. 2018). The hippocampus is already known to be critical for long-term memory formation (Andersen et al. 2007; Kandel, Schwartz, and Jessell 2000). The co-location of a navigational and memory system gives neural support to memory-enhancing techniques that involve picturing the sequence of memory items around one's living room or along a familiar trail. But what about other aspects of cognition, such as planning or reasoning in general?

One possibility involves the conceptual spaces (Gärdenfors 2000) discussed briefly in chapter 2 (and related to the hippocampus according to the Moser group (Bellmund et al. 2018)). Might grid cells provide a mechanism for systematically moving about such spaces, with many possible actions (and their intensities) replacing head-direction and velocity cells, respectively? The key gradient in this scenario is that of grid-bump movement with respect to different actions, and such derivative information might be manifest in the synaptic strengths of connections from premotor areas to the entorhinal cortex.

For example, consider the planning that a coach might do for a sports team. Typical spectra of prominence include (a) the health and physical condition of the team, (b) the confidence that the players have in their own physical and mental preparedness, and (c) the coherence and unity that the players exhibit in their style of play and attitudes toward one another. In reasoning about these factors and their interactions, each of these three spectra might link to a particular set of grid cells, as would neurons representing particular actions, such as those of the premotor cortex. Then, reasoning about the consequences of one such action (such as the grueling group sprints that most athletes detest) with respect to a spectrum (such as that of health-and-conditioning) would *drive the bump* along the relevant subnetwork of MEC.

Expecting these spectra to wrap around to form cycles (or a torus in a multidimensional context) may be unrealistic in some cases, but all three mentioned above have clear cyclic tendencies if driven by certain actions. Physical condition, for example, normally improves with exercise, but an excess can quickly produce injuries and, essentially, a wraparound from *excellent* to *poor*. The same holds for the confidence spectrum, wherein players can become increasingly positive about their abilities as they tackle more challenges, but one bad experience against a dominant opponent (i.e., one challenge too many) can undo a lot of progress if not properly addressed by the coach during and after the defeat. Finally, coherence may gradually improve with repetition of systematic plays among a stable group of players, but adding in new plays or players too quickly can move the needle straight back to chaos.

The coach's predictive reasoning might then go as follows. After assessing the current points on each of these spectra at the start of a season, the coach comes up with a training plan (a set of actions) designed to improve, or at least maintain, each of them. The current state would correspond to a location in an abstract space, and thus one or more related place cells (P*) in the hippocampus. These place cells would invoke a set of grid cells (G*) that best support them. Next, in contemplating the training plan, the coach invokes several potential actions (A*), including physical exercises and drills, team meetings, and the scheduling of early-season practice opponents.

Then, G* and A* interact to predict the team's future state. First, activation of A*-correlated neurons drives various MEC bumps from the G* state to a new grid state, G**. Next, G** maps to a new place-cell group, P**, which corresponds to the predicted state of the team. This *turn-the-MEC-crank* model of planning simply exapts the entorhinal and hippocampal predictive machinery that originally evolved for a purely spatiotemporal task, navigation. It relies heavily on an ability of the brain to map conceptual spaces (i.e., spectra) to grid cells and to incorporate a gradient-based understanding of the causal relationships between agent' actions and translations within these spectra.

In short, learning to make good predictions, of many varieties, may involve tuning of the synapses between the neo- and entorhinal cortices, and the hippocampus proper. The human use of spatiotemporal analogies and metaphors is well documented in mathematics (Lakoff and Nunez 2000) and in numerous other areas (Lakoff and Johnson 1980). This provides psychological evidence for a neural theory linking many cognitive abilities (including prediction) to the navigational apparatus of the hippocampal system. Although nothing has been proven, the possibility comes up frequently in research articles on grid and place cells.

3.5 Gradients of Predictions in the Basal Ganglia

The neural mechanisms for prediction and gradient calculation also come together in the basal ganglia, shown in figure 3.27. Roughly speaking, the basal ganglia receives inputs from the cortex via the striatum, whose neurons have firing properties amenable to *context detection*: they require a significant amount of cortical input in order to fire, so those cortical firing patterns constitute a context, which the basal ganglia then maps to two outputs: an action and an evaluation. Actions are (clearly) critical for proper behavior, while evaluations motivate the neural modifications that underlie learning.

The basal ganglia's action-evaluation separation closely matches a fundamental paradigm of reinforcement learning (RL) known as the *actor-critic model* (Sutton and Barto 2018), wherein an actor module handles action selection while its counterpart, the critic, handles evaluation. In RL, communication between the two is restricted to an error term (computed by the critic) that amounts to a temporal prediction gradient: the difference between predictions made at times t+1 and t.

Focusing on the critic segment of figure 3.27 (right), several pathways lead from the striatum to the substantia nigra pars compacta (SNc), which, by many accounts (Barto

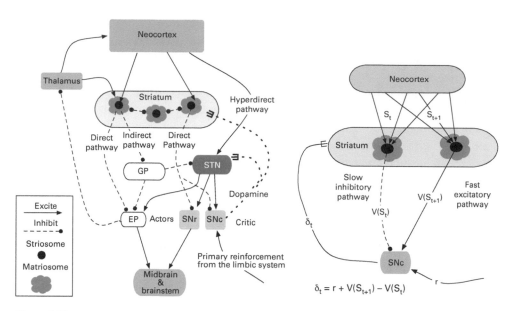

Figure 3.27
(Left) General functional anatomy of the mammalian basal ganglia. (Right) Computation of temporal-difference error (δ_t) by the SNc of the basal ganglia.

1995; Houk, Davis, and Beiser 1995; Graybiel and Saka 2004) constitutes the critic's core. Although the neuroscientific accounts vary, many models (Graybiel and Saka 2004; Prescott, Gurney, and Redgrave 2003) agree on the presence of both (a) fast excitatory connections from the neocortex and striatum to the SNc, and (b) slower inhibitory links between those same regions. Finally, one well-known input to the SNc comes from the amygdala, the center of the brain's emotional response (LeDoux 2002); it activates whenever the body experiences pleasure or pain.

Taken together, these three main pathways to the SNc (summarized in figure 3.27, right) form an ideal manifestation of RL's central prediction gradient, known as the *temporal difference error* (TDE). Essentially, TDE represents the difference between (a) the estimated value ($V(s_t)$) of the state/context(s) experienced by a system at time t, and (b) the combination of another estimate ($V(s_{t+1})$) for the system's next state, and any reward or penalty (r) incurred between times t and t+1. In RL, this is commonly expressed as

$$\delta_t = V(s_{t+1}) + r - V(s_t) \tag{3.2}$$

TDE (δ_t) is then combined with a learning rate (α) to update the estimate $V(s_t)$:

$$V(s_t) \leftarrow V(s_t) + \alpha \delta_t \tag{3.3}$$

This embodies *bootstrapping*, wherein an estimate made at time t is improved by one made at time t+1. Why should estimates made at t+1 be any more accurate than those at t? The answer revolves around the formal definition of these estimates: in RL theory, $V(s_t)$ embodies a *prediction* as to the amount of reward that the agent will accrue from time t until the end of that particular problem-solving attempt (formally known as an *episode*), often denoted as time T. It therefore makes sense that a prediction made at time t+1 and involving a fixed time horizon (T) should be marginally more accurate than one made at time t. In addition, the traversal from s_t to s_{t+1} involves concrete feedback (reward or punishment), r, which can only enhance the realism of $V(s_t)$. In short, the derivative of the prediction (supplemented with the immediate reward) provides the basis for an improvement of the prediction.

Returning to the basal ganglia, the analog of $V(s_t)$ is the activity level of the striatal neurons that have been tuned to detect s_t, which, via the fast excitatory pathway, correlates with the strength of positive inputs to SNc. Crucially, the SNc controls the diffuse secretion of the neuromodulator dopamine, whose presence enhances the synaptic modifications that enable learning; and the striatum receives a substantial amount of this dopamine, thereby affecting the synapses from cortex to striatum. Of special importance here is the well-documented fact (Schultz et al. 1992) that dopamine secretion results from a mismatch between expectations and reality, but not from the mere presence of a reward signal from the amygdala. In other words, when that reward is *expected*, the SNc remains inactive. In terms of the incoming connections to SNc in figure 3.27, a high value of r excites SNc only if not counterbalanced by a strong inhibitory signal from the striatum, where that signal represents $V(s_t)$, a prediction of reward.

Furthermore, SNc stimulation can occur in the absence of an amygdalar impulse. When the magnitude of the excitatory input from the fast pathway (which conveys $V(s_{t+1})$) exceeds that of the delayed inhibitory signal (for $V(s_t)$), this difference between two predictions can also activate SNc, resulting in dopamine and learning. In psychological terms, this difference represents a heightened anticipation of reward. In the context of this chapter, this difference signifies a positive prediction gradient that leads to learning, which works to reduce that gradient by increasing $V(s_t)$. As shown in the figure, the activity level of the SNc embodies the TD error of RL, via its relationship to $V(s_t)$, $V(s_{t+1})$ and r.

One final question involves timing. The SNc computes a form of TD error at time t+1 (or slightly thereafter) based on a fast excitatory value signal pertaining to s_{t+1} plus a delayed inhibitory signal representing $V(s_t)$. So how does the dopamine produced at time t+1 lead to an update of the synapses encoding $V(s_t)$ but not $V(s_{t+1})$? One plausible answer stems from a complex network of chemical interactions involving some inherent latencies (of approximately 100 milliseconds) (Houk, Adams, and Barto 1995; Downing 2009), the details of which are beyond the scope of this book. This intricate neurochemistry ensures that only those synapses (from neocortex to striatum) that have recently (but not too recently) been active are susceptible to the modifications induced by dopamine.

In summary, the basal ganglia calculates a temporal prediction gradient via complex neural circuitry that includes a crucial delayed inhibitory signal. The difference between a fast excitatory and delayed inhibitory signal (both representing the values of states in terms of their *predicted* future reward) combines with a fast excitatory indicator of *immediate* reward to yield a signal whose neurophysiological manifestation is dopamine and whose computational analog is the classic temporal-difference error of reinforcement learning theory. In both fields of study, that signal leads to learning of an improved prediction for the context associated with the delayed inhibitory signal: s_t.

Whereas navigating bacteria and our early levels of neural processing use gradients as the basis for a prediction, in higher levels of the brain, such as the basal ganglia, the key gradients are of the predictions themselves, and these gradients then govern tuning of the predictive machinery. This general trend continues up into the highest brain regions, where a great many signals probably represent abstractions and expectations rather than reality.

3.6   Procedural versus Declarative Prediction

In an earlier pair of journal articles, I analyzed several brain regions (neocortex, hippocampus, thalamus, cerebellum, and basal ganglia) with respect to procedural versus declarative knowledge (Downing 2007a) and then how these differentially facilitate prediction (Downing 2009). These same cranial regions receive considerable treatment in this and other chapters, although this book will not delve as deeply into those neural circuits as do the two articles. Here, the focus is more on the mathematical and computational components of predictive networks.

However, the distinction between procedural and declarative prediction deserves mention. In the former, an agent may act *as if* it had knowledge of a future state without actually having an internal representation of that expectation. For example, gradient-following bacteria exhibit procedurally predictive behavior. The cerebellum and basal ganglia often carry the *procedural* label, as their numerous parallel cables manifest relatively hardwired (though modifiable) links between contexts and actions. The sheer density of these connections equips the organism with many tunable if-then situation-action rules for survival, and many of these handle the temporal differences required for predictions (as detailed in Downing 2009). But they do not facilitate the formation of stable neural activation patterns suggestive of *representations*. Without stability, a pattern cannot persist long enough to form the basis of attention, which is critical for everything from simple conscious reasoning to the advanced use of symbols and language (Deacon 1998).

The hippocampus and neocortex (alone and in combination with the thalamus) have a different architecture, one lacking parallel lines but replete with recurrent connections; and these often provide the feedback necessary to produce stable activation patterns. When those patterns represent future states, they constitute declarative predictions. Similarly, they can represent declarative goals; and both predictions and goals, when encoded as activation patterns, can serve as inputs to comparators that combine their inverses with representations of reality. These differences can have downstream effects that make goals and predictions functionally significant. Forming and combining declarative representations provides much more computational flexibility than do behaviors embodied solely in situation-action associations. Thus, higher organisms, with more declarative capabilities (evidenced by expanded hippocampal and/or neocortical regions) are capable of more complex problem solving.

3.7   Rampant Expectations

The organisms, simple circuits, and complex brain regions above illustrate key relationships between several core concepts of this book: gradients, sums, predictions, and adaptation. Bacteria employ chemistry to compute gradients, which then influence behavior *as if* the organism has informed expectations about its immediate spatiotemporal future. The nematode worm displays similar implicit (procedurally predictive) behavior, but using basic desensitization mechanisms within its primitive nervous system. These behaviors exhibit procedural predictivity in the same way that a baseball outfielder *predicts* where a fly ball will land: he gradually moves to the proper location while tracking the ball but presumably has no internal mental representation of the final destination. Similarly, the outfielder uses gradient information (changes to the elevation angle of the ball) to adjust his own velocity and direction of movement. In all of these cases, the gradient enables prediction, albeit implicitly.

Conversely, in the mammalian visual system (and many other brain regions), neighborhood activity-level averages function as predictions of the activity level of any neuron (N), with the difference between N's activation and the prediction constituting a spatial gradient sent up the neural hierarchy. So in this case, the prediction enables gradient calculation, which, at the next level, supports further averaging and prediction. Note that the predictions themselves become more explicit in this context, since they are directly reflected in the total inhibition received by N. Here, the immediate result is not improved overt behavior but more broadband and energy-efficient signal transmission.

In the simple computational models based on Tripp and Eliasmith's (2010) neural motifs for temporal differentiation, explicit predictions stem from the leaky integration of rising temporal gradients. When projected back downward, these predictions normalize the lower layer's state value to form an error term. In chapter 5 on predictive coding, that error term is shown to serve as the main ascending signal. These models indicate that the interactions between gradients and integrative predictions support neural hierarchies in which higher layers can run at much slower speeds while still providing coarse, but reasonably accurate, predictions of lower-level behavior.

These simple circuits, whether alone or in repetitive layers, provide relatively simple, generic mechanisms for prediction that could potentially exist in many parts of the brain, particularly the neocortex, either as large fields of predictors or as small islands of expectation production. In contrast, several complex, heterogeneous brain areas appear to house potent predictive machinery of very specialized design, as seen in the cerebellum, hippocampus, and basal ganglia.

The convergence of prediction and control seems particularly evident in the cerebellum, a pivotal area for the timing and coordination of both motor and cognitive activities. Although this region is often modeled as an adaptive controller, it does not cleanly decompose into neural modules that map to regulator components. Instead, the numerous granular cells and emanating parallel fibers appear to transmit goals, predictions, and sensory reality, which then combine at comparators in the Purkinje cells to yield prediction errors, which then determine control signals sent to actuators, whose efference copies then provide predictions that feed back into the granular cells. Interactions of this control loop with a teaching signal from the inferior olive provide the brain's best example of adaptation via supervised learning (Doya 1999).

The hippocampus is particularly intriguing for its predictive potential. First, area CA1 has been proposed by neuroscientists as a comparator of reality signals—coming directly from the entorhinal cortex (EC)—and (slightly delayed) predictions based on pattern sparsification in area DG followed by pattern-completion of memories in area CA3. These memories serve as more abstract, slower-time-scale representations, whose comparison to more-immediate sensory reality indicates the level of surprise associated with the current situation. In addition, the interaction between the EC's grid cells and the hippocampal place cells has very strong implications for prediction, with the former updating the latter as to current and future locations of the agent in the absence of sufficient teleosensory information (e.g., sight, sound, smell) but based on transitions in collections of grid-cell networks as driven by proprioceptive velocity and orientation signals. Here, gradient knowledge, of how changes in position (i.e., velocity) affect activity-bump movement in grid cells, directly supports navigation by dead reckoning; and this same apparatus may be co-opted for numerous other cognitive activities (Bellmund et al. 2018).

Finally, the architecture and dynamics of the basal ganglia couple gradients, prediction, and adaptation. Employing signal-delay circuitry (similar to Tripp and Eliasmith's motifs) in combination with immediate reinforcement information from the amygdala, the basal ganglia compute predictions (of future reinforcement) at two adjacent time points, the difference of which manifests a temporal prediction gradient, which essentially represents *surprise*. Adaptation, via dopamine-enhanced synaptic modification to basal gangliar afferents, is then driven by the level of surprise.

As this chapter indicates, there are many neural structures capable of generating predictions. Some employ gradients and integrals in simple networks, while others exploit these same basic components in circuits resembling PID or adaptive controllers, thus blurring the boundaries between prediction and control. Reconciling these functional models with actual neural anatomy and physiology is always a speculative affair, but the entire field of computational neuroscience is built on a wide range of theories having supporting, but hardly confirming, evidence. But if prediction is indeed one of the brain's primary functions, then some of these predictive interpretations of contemporary neuroscience deserve further consideration and exploration.

4 Neural Energy Networks

4.1 Energetic Basis of Learning and Prediction

Herman von Helmholtz (1821–1894) was a nineteenth-century polymath whose achievements spanned thermodynamics, physiology, geometry, topology, and psychology (Patton 2018). One of his primary, lasting contributions to cognitive science was the distinction between sensations (the status of the peripheral sensory receptors) and percepts, realized as *mental adjustments*. He believed that many mental constructs were learned, not innate, thus bringing him into conflict with nativist theories of perception.

Helmholtz employed concrete mathematical principles in all of his work, whether theoretical or empirical, including his analysis of the brain. He viewed perception as a form of statistical inference in which mental constructs formed as hypotheses of the probable causes of sensations. This perspective motivated Peter Dayan and colleagues (1995) to design the Helmholtz machine, a neural network that takes a Bayesian statistical view of perception and enhances it with an *analysis-by-synthesis* approach to pattern recognition (summarized by Yuille and Kersten 2006), wherein the interpretation of sensory input improves dramatically when coupled with a generative model for producing expected sensory patterns from internal mental states (embodying causal explanations).

In the Helmholtz machine and related networks (as well as in the more theoretical work of Karl Friston described below), terms such as *explanation*, *cause*, and *causal explanation* have a fairly abstract, generic meaning that includes standard physical causality, as in gravity causing a ball to roll down an incline. However, much of the literature seems to view causes as the objects or events that lead to the sensory input of the agent. Thus, a spoon is the cause of the sensations in my hand when I eat soup, which, in turn, is the cause of the sensory patterns on my taste buds. In this way, causes can often be equated with classes in supervised learning.

A key premise of analysis-by-synthesis is that the space of causes of any sensory pattern is (for all intents and purposes) infinite. Thus, unidirectional inference from sensations to causes is bound to drown in a flood of plausible, but conflicting, causes, unless certain causes have a probabilistic bias that can be exerted in a top-down manner to favor some and inhibit other streams of bottom-up signals. The Helmholtz machine incorporates this bidirectional activity such that interleaved processes of interpretation and generation promote the gradual emergence of a tight coupling between the neural system and the environment (via sensory receptors), with this coupling defined statistically as high similarity (low divergence) between two probability distributions over the causal states: one produced by the recognition machinery, and the other by the generative mechanisms.

Herman von Helmholtz's additional posthumous contribution to the Helmholtz machine is Helmholtz free energy, a thermodynamic concept easily retooled for machine learning to capture relationships between the above-mentioned probability distributions, with the link between energy and probability coming from another primitive of thermodynamics: the Boltzmann equation. This interchangeable currency of *enerbility* (coined here, solely for the purpose of this chapter) provides an appropriate metric for the emergent progress of Helmholtz machines (and related networks) as they cycle through recognition and generative phases.

This learning progress fully earns the adjective *emergent*, since enerbility metrics (such as free energy) apply to the global state of a network but can fortuitously be transformed into local, Hebbian learning rules by calculating the gradients of these metrics with respect to individual synaptic weights. Thus, an iterative top-down, bottom-up cycle partnered with Hebbian parameter updates can produce a neural system that is well-tuned to its environment, and all without any form of supervisory feedback.

4.2   Energy Landscapes and Gradients

Although neural systems have been thoroughly analyzed from physicochemical energetic perspectives (Sterling and Laughlin 2015; Stone 2018), a more abstract concept of energy, borrowed from spin-glass theory, provides a popular and powerful metric for both assessing the behavior of neural networks and guiding their adaptation. It also serves as a starting point for more complex metrics that link energy and enerbility to prediction.

In 1982, John Hopfield made the seminal connection between spin-glass theory and neural networks (Hopfield 1982), thus lending a bit of hard-science legitimacy to a field, connectionism, that was struggling for respect, prior to convincing demonstrations of the backpropagation algorithm's utility in 1986 (Rumelhart, Hinton, and Williams 1986). In this interpretation, neural network energy embodies *conflict* between the activation levels of neuron pairs and the sign and magnitude of the synaptic link between the two units. As sketched in figure 4.1, four primary motifs illustrate the general relationship: when the activity levels of paired neurons match the synaptic type, the energy of that pairing is low. Otherwise, an inverse correlation signals high energy. For example, when both pre- and postsynaptic neurons exhibit high activity, and the synapse between them is excitatory (i.e., has a positive weight / strength), the local energy of that pairing is low: no conflict. Anthropomorphically speaking, an excitatory synapse *wants* high postsynaptic activity as a consequence of presynaptic vigor, while an inhibitory synapse *wants* presynaptic activity to reduce postsynaptic firing. Low energy indicates that synapses are getting what they want. High energy reflects their dissatisfaction.

In the original Hopfield network, all connections are bidirectional, with a single weight (strength), so the pre-versus-postsynaptic distinction vanishes, but the basic correlation between the paired activity levels can still be compared to the weight's sign to assess the level of conflict. In either bidirectional or unidirectional networks, total energy is simply

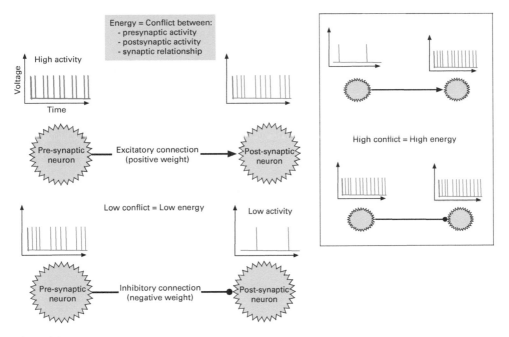

Figure 4.1
Four different scenarios showing the parallel between spin-glass-style energy and the relationships between two neurons and their connecting synapse. (Main) Low energy reflected in a match (no conflict) between paired neurons and synaptic predisposition (i.e., excitatory or inhibitory). (Inset) High energy stemming from high conflict between paired activity and synaptic type.

the sum of all local energies: the sum of all energies from all pairs of directly connected neurons. Most Hopfield networks are fully connected: every neuron synapses with every other unit.

In the following analysis, the outputs (x) of neurons are binary: either -1 or 1,[1] and synapses are presumed bidirectional, with real-valued (positive or negative) weights (w). Then $-x_j x_k w_{jk}$ expresses the energy at a particular synapse, with the energy of the entire network (E) given by

$$E = -c_1 \sum_{j,k} x_j x_k w_{jk} - c_2 \sum_k x_k x_k^* \tag{4.1}$$

where c_1 and c_2 are positive constants, and x_k^* represents the value loaded into unit k at the start of a run. The relationship between x_k's current value and its initial value can also serve as a source of conflict, but the following analysis ignores this second term of equation 4.1.

Hopfield networks exhibit adaptation in both the short and long term, with both processes working to reduce E. In the short term, the weights remain fixed while the activation levels change. In the long term, learning occurs via weight change. Both modifications stem from gradients of E.

Hopfield nets are popular theoretical and educational tools but have little practical value. Their weights can be tuned (via equation 4.5) to provide imperfect storage for a limited number of patterns, which can then be *recalled* by feeding pattern fragments into the net and letting it run to equilibrium; the output core state then serves as a reconstruction of the original pattern. Recall errors occur frequently when the original patterns are too plentiful or too similar to one another. This disappoints many machine-learning researchers but intrigues cognitive scientists, who appreciate the relationships between remembering, forgetting, and abstracting.

The significance of Hopfield nets for this chapter lies in the relationships among neural activity, synaptic disposition, and energy, along with the biologically realistic mechanisms by which global energy is reduced by *purely local*, gradient-based operations. They motivated more advanced, but equally local and energy-based, models invented afterward.

Unit Activation and Hebbian Learning in the Hopfield Network

In the Hopfield net, short-term modification involves changes to the activation levels of units based on their sum of weighted inputs, as in most neural networks. However, in this case, those changes directly help satisfy the objective function: they reduce E. To see this, compute the derivative of E with respect to the activation level of any unit, x_a:

$$\frac{\partial E}{\partial x_a} = \frac{\partial}{\partial x_a}[-c_1 \sum_{j,k} x_j x_k w_{jk} - c_2 \sum_k x_k x_k^*] = -c_1 \sum_{j \neq a} x_j w_{ja} - c_2 x_a^* \tag{4.2}$$

This relationship assumes a bidirectional Hopfield net in which the weights are labeled such that all weights connected to x_a use a as the second subscript. To reduce error (E), take the negative of this derivative when updating x_a:

$$x_a = -\frac{\partial E}{\partial x_a} = c_1 \sum_{j \neq a} x_j w_{ja} + c_2 x_a^* \tag{4.3}$$

Note that this is just the sum of weighted inputs to x_a plus a bias associated with its initial value, x_a^*. Finally, convert x_a to a -1 or $+1$ depending on its sign (with zero also mapping to -1). In the normal operation of a Hopfield network, initial values are placed on each unit (as shown in figure 4.2) and then neurons are randomly and asynchronously chosen to update their activation values based on equation 4.3, with each update reducing the local contribution to E. Over time, the network settles into a stable configuration (i.e., no neurons change state (-1 or 1) after updating) which tends to have low global energy (E).

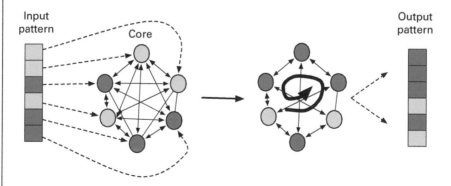

Figure 4.2
Basic topology and operation of a Hopfield network, where inputs load directly onto the core neurons. After many rounds of asynchronous updating, the core reaches equilibrium and its activation levels constitute the net's output.

Hopfield networks are designed to hold many patterns (P) at once, with each pattern distributed over all neurons. This type of knowledge (memory) resides in the weights, not the individual activations, and these weights represent general correlations between pairs of neurons as calculated over all of P. Weight modifications based on any single pattern must therefore be scaled (by a learning rate, λ) such that each pattern contributes to but does not dominate the final weights. For each pattern (p), the weight update again stems from the derivative of E (this time, with respect to the weight):

$$\frac{\partial E^p}{\partial w_{ab}} = \frac{\partial}{\partial w_{ab}}[-c_1 \sum_{j,k} x_j^p x_k^p w_{jk} - c_2 \sum_k x_k^p x_k^{p*}] = -c_1 x_a^p x_b^p \tag{4.4}$$

The weight change is then

$$\triangle w_{ab} = -\sum_{p \in P} \lambda \frac{\partial E^p}{\partial w_{ab}} = -\lambda \sum_{p \in P} -c_1 x_a^p x_b^p = \lambda \sum_{p \in P} c_1 x_a^p x_b^p \tag{4.5}$$

This constitutes a simple, local, Hebbian update, based on the activation levels of the neurons on each end of the synapse.

4.3 The Boltzmann Machine

Introduced in 1985 by Ackley, Hinton, and Sejnowski, the connectionist version of the Boltzmann machine (BM) exploited the critical link between energy and probability (expressed in the Boltzmann equation) to provide further cognitive realism to the Hopfield network. The underlying philosophy of these networks is that proper understanding and interpretation of (input) data can only be achieved by systems capable of *generating* such data (as output). Their Boltzmann machines (Ackley, Hinton, and Sejnowski 1985) therefore perform the complementary tasks of recognition and generation using (crucially) the same neural substrate: the deep understanding resides in neurons and synapses involved in both processes, and that comprehension arises precisely because of this coupling, via a mechanism known as *contrastive divergence*, a purely local mixture of Hebbian and anti-Hebbian operations.

Linking Probability and Energy in Boltzmann Machines

The key starting point is the Boltzmann distribution from statistical mechanics:

$$p(s_i) = \frac{e^{\frac{-E(s_i)}{kT}}}{\sum_j e^{\frac{-E(s_j)}{kT}}}$$

where $E(s_i)$ is the energy of the ith system / network state, $p(s_i)$ is the probability of the system occupying that state, T is the temperature (Kelvin), and k is the Boltzmann constant. The denominator, known as the *partition function*, is often symbolized by Z:

$$Z = \sum_j e^{\frac{-E(s_j)}{kT}} \tag{4.6}$$

The Boltzmann distribution expresses probability as a function of energy, with high-energy states being less likely than low-energy situations, just as observed in the physical world. In artificial intelligence and information theory, analogs of the Boltzmann equation traditionally ignore the Boltzmann constant, and often the temperature as well, leaving the following abstraction used throughout the remainder of this chapter:

$$p(s_i) = \frac{e^{-E(s_i)}}{Z} \qquad (4.7)$$

where $Z = \sum_j e^{-E(s_j)}$. The inverse will also assist in several later calculations:

$$E(s_i) = -ln[p(s_i)Z] \qquad (4.8)$$

A wide variety of network topologies can function as Boltzmann machines, as shown in figure 4.3. Most descriptions in the literature, including the original article (Ackley, Hinton, and Sejnowski 1985), involve bidirectional weights and a fully intraconnected core (aka hidden layer), but the original article also includes an example network similar to that in the lower left of figure 4.3. The neurons normally have binary activation states that are

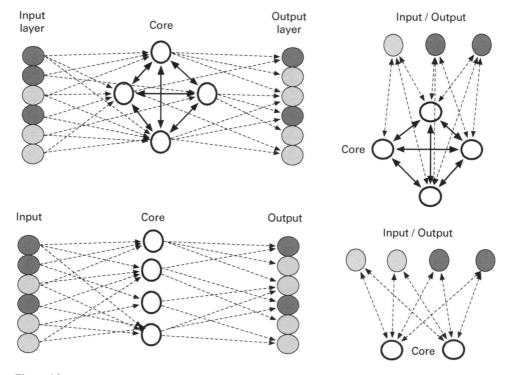

Figure 4.3
Assorted Boltzmann machine topologies, which vary in terms of the connectivity of the core / hidden units, differentiation between input and output neurons, and connectivity (unidirectional or bidirectional) between the core and inputs / outputs.

determined stochastically based on the sum of weighted inputs (S) passed through a sigmoid function, $f(S) = \frac{1}{1+e^{-S}}$, which yields the probability of the neuron outputting +1.

The key theoretical contributions of the Boltzmann machine are easiest to understand using a topology similar to that in the upper right of figure 4.3, where inputs and outputs share the same interface units, which are fully connected to the core. When given an input pattern, the network runs to equilibrium, at which point the state of the core neurons constitutes an *explanation* of the input. By restarting the network with an empty interface layer but with that explanation loaded into the core, the network should eventually produce the original input on the interface. The explanation thereby serves as both an interpretation and generator of the input.

When the environment provides data vectors $d \in D$ to a Boltzmann machine, it should learn to produce explanations for each. Thus, the BM modifies its structure (i.e., weights) to reflect or *represent* D. In response to any $d \in D$ clamped to its input units, the BM's core should transition to equilibrium state s_d, which should be a local minimum on the energy landscape. Conversely, when presented with an input pattern not in D, the BM should transition to a high-energy equilibrium. Figure 4.4 portrays an energy landscape before and after training, showing how learning (via weight change) contorts the terrain such that states that explain members of D occupy low-energy basins, while those explaining nonmembers sit atop bulges. This combination of valley and ridge formation creates ample separation between members and nonmembers, thus producing differences (in energy) that make a difference (in class membership). Energy in Boltzmann machines is typically expressed as conflict between paired neurons and their connecting weight, similar to equation 4.1 for Hopfield nets.

However, whereas the dynamics of the Hopfield nets are designed to move core states toward low energy, the BM's overriding goals are statistical: it adapts to generate a probability distribution over its output vectors that matches the frequencies of $d \in D$. And, as seen in equation 4.7, the probability of a state is a function of its energy, but also of the partition function (Z), which plays a very significant role. In effect, the energy drives valley formation, while Z affects ridge formation in the energy landscape; it all falls out of a few straightforward calculations that underlie contrastive Hebbian learning (as explained in the accompanying box, "Contrastive Hebbian Learning (CHL)").

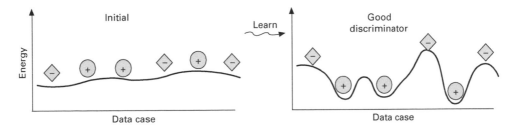

Figure 4.4
The energy landscape, a mapping from environmental inputs, aka data (D), to the energies inherent in the neural-network states that evolve to *explain* the data. The positive (circle) cases represent $d \in D$, while negatives (diamonds) represent $d \notin D$, that is, cases that the system would not experience in its normal environment but is still capable of processing during testing. Learning deforms the landscape such that members engender low-energy states, while nonmembers promote high-energy explanations.

Contrastive Hebbian Learning (CHL)

The basis of CHL is the relationship between weight change and the change in probability of a system state, $\frac{\partial p(s_i)}{\partial w_{jk}}$. Since the expression for $p(s_i)$ in equation 4.7 is the quotient of two terms involving energy, it is advantageous to use the natural logarithm of $p(s_i)$. Exploiting the positive monotonic relationship between $p(s_i)$ and $ln(p_si)$ is particularly useful in situations such as this, where the goal is to optimize $p(s_i)$ by changing w_{jk}.

$$ln(p(s_i)) = ln(\frac{e^{-E(s_i)}}{Z}) = ln(e^{-E(s_i)}) - ln(Z) = -E(s_i) - ln(Z) \tag{4.9}$$

Then:

$$\frac{\partial ln(p(s_i))}{\partial w_{jk}} = -\frac{\partial E(s_i)}{\partial w_{jk}} - \frac{\partial ln(Z)}{\partial w_{jk}} \tag{4.10}$$

Using equation 4.4 for the derivative of the energy with respect to any weight

$$\frac{\partial ln(p(s_i))}{\partial w_{jk}} = x_j^{(i)} x_k^{(i)} - \frac{\partial ln(Z)}{\partial w_{jk}} \tag{4.11}$$

Since $\partial ln(f) = \frac{\partial f}{f}$:

$$\frac{\partial ln(Z)}{\partial w_{jk}} = \frac{\frac{\partial Z}{\partial w_{jk}}}{Z} = \frac{1}{Z}\frac{\partial Z}{\partial w_{jk}} = \frac{1}{Z}\frac{\partial(\sum_a e^{-E(s_a)})}{\partial w_{jk}} = \tag{4.12}$$

$$\frac{1}{Z}\sum_a \frac{\partial e^{-E(s_a)}}{\partial w_{jk}} = \sum_a \frac{e^{-E(s_a)}}{Z}\frac{-\partial E(s_a)}{\partial w_{j,k}} \tag{4.13}$$

Using equations 4.7 and 4.4, the derivative for $ln(Z)$ becomes

$$\frac{\partial ln(Z)}{\partial w_{jk}} = \sum_a p(s_a)x_j^{(a)} x_k^{(a)} \tag{4.14}$$

This yields the final expression of equation 4.15, which consists of a Hebbian term involving the values of x_j and x_k in the current state, s_i, and an anti-Hebbian term involving *all possible* states and their associated x_j and x_k values:

$$\frac{\partial ln(p(s_i))}{\partial w_{j,k}} = x_j^{(i)} x_k^{(i)} - \sum_a p(s_a)x_j^{(a)} x_k^{(a)} \tag{4.15}$$

Notice that the partial derivative of Z (from equation 4.10) creates the large (often intractable) summation on the right of equation 4.15. Boltzmann machines and similar techniques often handle this problem via sampling to get reasonable estimates of the complete summation (as discussed below).

As mentioned earlier, the BM's goal is to match its output pattern distribution to all of D, not just to a single $d \in D$. This engenders either a maximization or a minimization problem: maximizing the log likelihood of generating D, or minimizing the divergence between the generated and target distributions. More formally, one approach seeks to maximize the average of $ln(p_g(s_d))$ over all $d \in D$: maximize $\langle ln(p_g(s_d))\rangle_{d \in D}$. The other attempts to minimize $D_{KL}(Q_D^0 \| Q_D^g)$, where Q_D^0 and Q_D^g denote the target and generated distributions, respectively. In both cases, the optimization problem relies on the derivatives of these objective functions with respect to individual network weights.

Beginning with maximization, for any weight w_{jk}, find its average effect over all $d \in D$ (where $\|D\| = N$):

$$\frac{\partial \langle ln(p_g(s_d)) \rangle_{d \in D}}{\partial w_{j,k}} = \frac{1}{N} \sum_{d \in D} \frac{\partial ln(p_g(s_d))}{\partial w_{j,k}} = \frac{1}{N} \sum_{d \in D} \{ \underbrace{x_j^{(d)} x_k^{(d)}}_{Hebbian} - \underbrace{\sum_a p_g(s_a) x_j^{(a)} x_k^{(a)}}_{anti\text{-}Hebbian} \} \quad (4.16)$$

Since the anti-Hebbian term is the same for each $d \in D$,

$$= \frac{1}{N} \sum_{d \in D} x_j^{(d)} x_k^{(d)} - \sum_a p_g(s_a) x_j^{(a)} x_k^{(a)} \quad (4.17)$$

The final result involves two different samples, over D and over S, the entire space. Abbreviate this result as δ.

$$\frac{\partial \langle ln(p_g(s_d)) \rangle_{d \in D}}{\partial w_{jk}} = \langle x_j^{(d)} x_k^{(d)} \rangle_{d \in D} - \langle x_j^{(a)} x_k^{(a)} \rangle_{a \in S} = \delta \quad (4.18)$$

Figure 4.5 illustrates CHL, wherein data from the environment drives a recognition phase in which the core units converge to an explanation state, s_d. Each such s_d for each $d \in D$ then provides data for the Hebbian updates of each of the network's weights (w): the activation levels (in s_d) of w's adjacent units. Conversely, during generation, a randomly initialized core drives the transition to equilibrium state s_a, whose activation levels then guide weight change in an anti-Hebbian manner.

A Boltzmann machine trained with CHL gradually learns to generate output distributions similar to D, even though the generation phase may experience only a small fraction of all $s_a \in S$. The cumulative Hebbian and anti-Hebbian effects still decrease KL divergence between these two distributions. Figure 4.6 depicts the dueling effects of Hebbian and anti-Hebbian learning, where the former depresses the energy landscape around the explanation states, while the latter elevates areas containing freely generated (aka *dream*) states.

Contrastive Hebbian Learning (Continued)

For the minimization problem, begin by elaborating the KL divergence:

$$D_{KL}(Q_D^0 \| Q_D^g) = \sum_{d \in D} p(s_d) ln \left(\frac{p(s_d)}{p_g(s_d)} \right) = \sum_{d \in D} p(s_d) ln(p(s_d)) - \sum_{d \in D} p(s_d) ln(p_g(s_d)) \quad (4.19)$$

This consists of an entropy and a cross-entropy term:

$$= -H(Q_D^0) - \langle ln(p_g(s_d)) \rangle_{d \in D} \quad (4.20)$$

Compute the derivative of these two terms with respect to any weight, w_{jk}, and note that the first term vanishes due to the independence of the target data distribution from w_{jk}:

$$\frac{\partial D_{KL}(Q_D^0 \| Q_D^g)}{\partial w_{jk}} = \frac{-\partial H(Q_D^0)}{\partial w_{jk}} - \frac{\partial \langle ln(p_g(s_d)) \rangle_{d \in D}}{\partial w_{jk}} = 0 - \frac{\partial \langle ln(p_g(s_d)) \rangle_{d \in D}}{\partial w_{jk}} = -\delta \quad (4.21)$$

Comparing equations 4.18 and 4.21 thus reveals

$$\frac{\partial D_{KL}(Q_D^0 \| Q_D^g)}{\partial w_{jk}} = -\frac{\partial \langle ln(p_g(s_d)) \rangle_{d \in D}}{\partial w_{jk}} \qquad (4.22)$$

Hence, whether maximizing the log likelihood ($\triangle w_{jk} = \lambda \delta$) or minimizing the KL divergence ($\triangle w_{jk} = -\lambda(-\delta)$), the prescribed weight modification is the same (where λ is the learning rate):

$$\triangle w_{jk} = \lambda \delta = \lambda \left(\langle x_j^{(d)} x_k^{(d)} \rangle_{d \in D} - \langle x_j^{(a)} x_k^{(a)} \rangle_{a \in S} \right) \qquad (4.23)$$

In the normal operation of a neural network, subsets of D (D', aka minibatches) are run through the network, and only a subset (S') of all states S will be sampled, so the more realistic weight-update rule becomes

$$\triangle w_{jk} = \lambda \left(\overbrace{\langle x_j^{(d)} x_k^{(d)} \rangle_{d \in D'}}^{\oplus} - \underbrace{\langle x_j^{(a)} x_k^{(a)} \rangle_{a \in S'}}_{\ominus} \right) \qquad (4.24)$$

Equation 4.24 forms the cornerstone of contrastive Hebbian learning (CHL) (Hinton et al. 1995; Hinton 2002; O'Reilly and Munakata 2000) in which weight updates stem from both Hebbian learning during a clamped recognition (aka wake) phase ($\oplus$) and anti-Hebbian learning during an unconstrained generative (aka predictive or dreaming) phase ($\ominus$).

The full details of the Boltzmann machine vary within the connectionist literature, and several nuances have been glossed over in the above description, including the simulated annealing process governing the transition to equilibrium states. The main concerns of this chapter are the relationships between probability and energy embodied in these networks (as defined by the Boltzmann equation), the relationships between probability distributions and energy landscapes, and the elegant fact that global optimization criteria can boil down to simple, local, Hebbian update rules for network weights. Furthermore, BMs give an early, primitive indication of the interactions between recognition and generation (aka parse and predict) and how this coupling yields deeper understanding than either process alone. Finally, contrastive Hebbian learning quantifies (weight) adaptation based on the difference between reality-driven and prediction-based system states, a recurring theme in this chapter.

4.4   The Restricted Boltzmann Machine (RBM)

Boltzmann machines serve a more theoretical than practical purpose because of the computational demands of simulated annealing across a large neuronal population: it can take a long time to reach equilibrium. However, by retaining the basic philosophy of the Boltzmann machine but modifying the topology, Hinton and Salakhutdinov (2006) transformed connectionist networks from interesting cognitive science models to powerful engineering tools, thus ushering in the field of deep learning in the first decade of the twenty-first century.

Based on Smolensky's harmony theory (1986), and similar to networks based on adaptive resonance theory (Carpenter and Grossberg 2003), the restricted Boltzmann machine

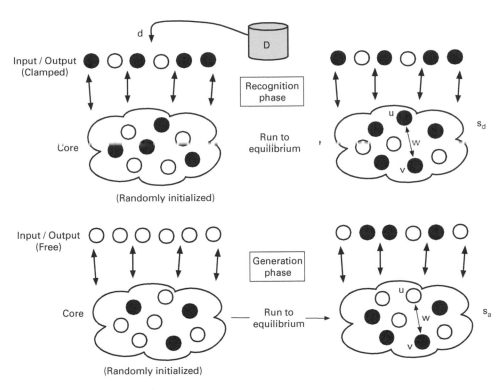

Figure 4.5
The two phases of contrastive Hebbian learning (CHL). (Top) Recognition: with environmental inputs $d \in D$ imposed on the interface neurons, the BM runs to equilibrium state s_d, in which the update to each weight w is $\triangle w = \lambda uv$ for the adjacent units, u and v. (Bottom) Generation: with a randomly initialized core and no environmental forcing, the BM runs to equilibrium state s_a, in which the weight update is anti-Hebbian: $\triangle w = -\lambda uv$.

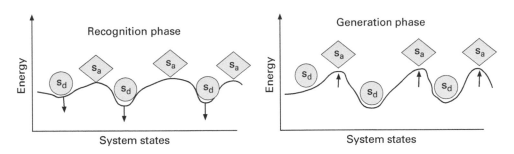

Figure 4.6
Modifications to the energy landscape incurred by contrastive Hebbian learning (CHL), where s_d denotes explanation states for $d \in D$ achieved during the recognition phase, while s_a signifies states produced by free / unclamped activity during the generative phase.

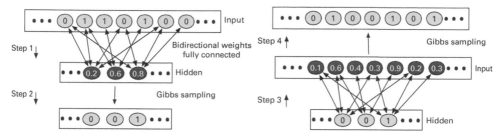

Figure 4.7
Two phases of the restricted Boltzmann machine (RBM) in one pair of layers. (Left) Recognition (aka *wake*) phase, where binary inputs drive real-valued activation levels of hidden-level neurons, which then serve as probabilities for Gibbs sampling to produce a binary vector. (Right) Prediction (aka *sleep*) phase, where binary hidden-level values determine real-valued input-level neurons, whose Gibbs sampling yields a *dream* input vector, which is then used at the start of the next recognition phase.

(Hinton and Salakhutdinov 2006) employs alternating bottom-up (recognition) and top-down (prediction) operations to gradually achieve harmony / resonance between two adjacent layers of a network. Once a low-level pair of layers harmonizes, the process can continue in stepwise fashion up through a deep neural hierarchy to eventually converge on system states that embody a deep, multileveled *understanding* of the environmental data.

The RBM consists of several stacked layers,[2] each of which fully connects to its neighbors (with bidirectional weights) but has no intralayer connections. By removing these intralayer links from the original BM design, the RBM simplifies the transition to equilibrium. In fact, Hinton (2002) invented contrastive divergence (CD)—a variant of contrastive Hebbian learning (CHL)—to preclude the equilibration process entirely and still guide weight change that reduces global energy and improves pattern recognition and generation.

Illustrated in figure 4.7, the RBM combines recognition and prediction phases, with each supplying a starting binary vector for the other. The mathematics of CD (Hinton 2002) verifies that even after one or a few rounds of this back-and-forth cycling between the two layers, the average activation levels of their neurons during the cycle(s) support appropriate Hebbian weight updates using the CD learning rule of equation 4.25:

$$\triangle w_{jk} = \lambda \left[\langle u_j v_k \rangle_{recognition} - \langle u_j v_k \rangle_{prediction} \right] \tag{4.25}$$

where $\langle u_j v_k \rangle_q$ denotes the fraction of times, averaged over all q (where q is recognition or prediction) phases of the current cycle(s), in which the jth neuron of the lower layer and the kth neuron of its upper neighbor layer are both on. Thus, learning in the RBM follows the similar pattern of Hebbian change based on recognition combined with anti-Hebbian influences from prediction.

After a pair of adjacent layers (L_k and L_{k+1}) has fully processed a set of input patterns (and updated the interlayer weights accordingly), the transition to the next pair (L_{k+1} and L_{k+2}) begins by mapping all of the original inputs to L_k through one recognition phase, yielding a set of activations for L_{k+1}, which then becomes the input set (i.e., targets) for the L_{k+1}-L_{k+2} cycles.

This incremental process proceeds upward through all layers of the RBM, thus achieving unsupervised adjustment of the network's weights, which then bias the RBM toward activation states that *explain* the environmental data (i.e., the inputs to layer L_0). To achieve competitive performance as a classifier, Hinton and Salakhutdinov added a single classifier head atop the RBM stack and then performed supervised learning across all layers, thus fine-tuning the weights for pigeonholing the environmental inputs into discrete classes. Prior to this work, networks with more than one or two layers were easy enough to train, since the backpropagation algorithm theoretically handles any number of layers, but performance of the resulting classifiers was very poor for these deep nets. The RBM broke through that barrier, as unsupervised pretraining massaged network weights into strongly biased starting points for supervised learning, and the benchmark results beat all competitors of that era. In the conclusion of their groundbreaking article, Hinton and Salakhutdinov (2006) remark: "It has been obvious since the 1980s that backpropagation through deep autoencoders would be very effective for nonlinear dimensionality reduction, provided that computers were fast enough, data sets were big enough, and the initial weights were close enough to a good solution. All three conditions are now satisfied."

Although the ensuing years have brought many breakthroughs in deep learning (as summarized by several researchers: Goodfellow, Bengio, and Courville 2016; Stone 2020; LeCun, Bengio, and Hinton 2015), thus surpassing the RBM and unsupervised pretraining as a favored machine-learning paradigm, these newer methods typically involve complex gradients (i.e., derivatives of loss functions with respect to weights) that display none of the locality or Hebbian nature of the Hopfield, BM, and RBM update rules. Each gradient's complexity stems from the long string of mathematical relationships required to link a weight to the loss. Although many of these networks have interesting biological parallels, they stray quite far from neuroscientific principles. Hence, they provide few computational insights into relationships among adaptation, prediction, and natural intelligence. As pointed out in LeCun, Bengio, and Hinton (2015), the vast majority of human learning is unsupervised, so why should we expect our top machine-learning methods to be supervised?

4.5 Free Energy

The above analysis of the Hopfield and Boltzmann machines essentially considers system states (s_d) as atomic units that include the data (d) that induces them; and, although no one-to-one mapping between d and induced states is assumed, both forms of networks have symmetric weights and run to equilibrium, thus limiting the number of attractors that the system settles into. However, in more complex networks, particularly those with asymmetric weights, the equilibria may be very hard to find, and hence the system may visit myriad *explanatory* states, with varying frequencies, all of which have different probabilities of generating d (see the accompanying box below for details).

Minimizing Free Energy

To understand how free energy applies to neural networks, begin by computing the probability of any data point (d) by marginalization across all possible causal states (s): $p(d) = \sum_s p(s, d)$. The posterior probability of explanation s′ given data d is then

$$p_g(s'|d) = \frac{p_g(s', d)}{p_g(d)} = \frac{p_g(s', d)}{\sum_s p_g(s, d)} \qquad (4.26)$$

where p_g are probabilities based on the activities of a neural network with weights g.

Given data d and an explanation state s', define the *energy of explanation* of the s'-d coupling as the *surprisal* (the negative log probability):

$$E_g(s';d) = -ln(p_g(s',d)) \tag{4.27}$$

Here, the semicolon in $E_g(s;d)$ indicates that E_g is a function of s, while d is fixed. Solving for $p_g(s',d)$:

$$p_g(s',d) = e^{-E_g(s';d)} \tag{4.28}$$

Dividing both sides by the marginal probability $p_g(d) = \sum_s p_g(s,d)$:

$$\frac{p_g(s',d)}{p_g(d)} = \frac{e^{-E_g(s';d)}}{\sum_s p_g(s,d)} \tag{4.29}$$

Combining equations 4.26 and 4.29 with the definition of the Boltzmann partition function, Z (described earlier):

$$p_g(s'|d) = \frac{e^{-E_g(s';d)}}{\sum_s e^{-E_g(s';d)}} = \frac{e^{-E_g(s';d)}}{Z_d} \tag{4.30}$$

From equations 4.29 and 4.30, clearly $Z_d = \sum_s p(s,d) = p_g(d)$, that is, the marginal probability of producing data point d over all neural network states $s \in S$. The additional subscript indicates the dependence of Z on data point d.

Equation 4.30 is nearly the same as equation 4.7, but this enhanced version separates the explanation (s') from the data (d) and quantifies the latter's probability of evoking the former via network operation.

As before, the primary goal of the network is to generate an output distribution of data commensurate with the environmental data that it perceives. If p(D) is the distribution over environmental inputs D, and $p_g(D)$ is the net-generated distribution of those same D vectors, then a standard level of generator success is the KL divergence between those two distributions:

$$D_{KL}(p(D),p_g(D)) = \sum_d p(d)ln\left(\frac{p(d)}{p_g(d)}\right) = \sum_d p(d)ln(p(d)) - \sum_d p(d)ln(p_g(d)) \tag{4.31}$$

Recognizing the first term as the negative entropy of p(D) and the second as the cross-entropy of p(D) and $p_g(D)$, this simplifies to

$$D_{KL}(p(D),p_g(D)) = -H(p(D)) + H(p(D),p_g(D)) \tag{4.32}$$

For later convenience, and as a reminder that the weighting of the negative logarithms comes from p, rewrite the cross entropy as

$$H(p(D),p_g(D)) = \langle -ln(p_g(D))\rangle_p \tag{4.33}$$

Then,

$$D_{KL}(p(D),p_g(D)) = -H(p(D)) + \langle -ln(p_g(D))\rangle_p \tag{4.34}$$

Given the goal of minimizing $D_{KL}(p(D),p_g(D))$ by adjusting the network's parameters (g), the entropy H(p(D)) can be ignored, since p(D) is independent of the network. The focus moves to minimizing $\langle -ln(p_g(D))\rangle_p$, and since p(D) is presumably a fixed distribution, the goal can be simplified to that of modifying g to minimize the sum of the negative log probabilities of each data point:

$$\sum_{d \in D} -ln[p_g(d)] \tag{4.35}$$

Ignoring the interactions between $p_g(d_1)$ and $p_g(d_2)$ for any $d_1, d_2 \in D$, and thus employing a divide-and-conquer strategy, the goal becomes one of reducing each of the negative log probabilities as much as possible: increasing the likelihood of generating the data. In machine learning, this term, $-ln[p_g(d)]$ is known as the *free energy* of the system with respect to d, $F_g(d)$, and it has interesting parallels to the physical concept.

In thermodynamics, the Helmholtz free energy (F) of a system is defined as its energy minus the product of its temperature (T) and entropy (H). It essentially measures the amount of energy that is available to do work; entropy measures disorder, which reduces the capacity for work.

$$F = \langle E \rangle - TH$$

where $\langle E \rangle$ denotes the average energy over all states that the unperturbed system may visit. Entropy depends on the probability distribution over those states.

In connectionism, free energy has a similar definition, with temperature often ignored:

$$F_g(d) = \langle E_g(s; d) \rangle_g - H_g(s|d) = -ln(p_g(d)) = -lnZ_d \tag{4.36}$$

In the definition of equation 4.36, a system experiences environmental input d and then moves through different internal states (s), with the visitation frequency determined by the parameters of the system (g). The energy (E_g) is therefore averaged over those states, that is, explanations. The entropy (H_g) reflects the distribution of those states and is therefore also a function of g. The equivalence of free energy and $-ln(p_g(d))$ is important and requires a short derivation (in the accompanying box below).

Equivalence of Free Energy and Negative Log Probability of Data

Since $\sum_s p_g(s|d) = 1$ over all states $s \in S$:

$$-ln(p_g(d)) = -\sum_s p_g(s|d)ln(p_g(d)) \tag{4.37}$$

Since $p_g(d) = \frac{p_g(s,d)}{p_g(s|d)}$:

$$-ln(p_g(d)) = -\sum_s p_g(s|d)ln\left(\frac{p_g(s, d)}{p_g(s|d)}\right) \tag{4.38}$$

Since $ln(x/y) = ln(x) - ln(y)$:

$$= -\sum_s p_g(s|d)ln(p_g(s,d)) + \sum_s p_g(s|d)ln(p_g(s|d)) \tag{4.39}$$

The rightmost term constitutes entropy:

$$= \sum_{s} p_g(s|d)(-ln(p_g(s,d))) - H_g(s|d) \tag{4.40}$$

Since $E_g(s; d) = -ln(p_g(s,d))$ and the summation is over all $s \in S$ weighted by $p_g(s|d)$:

$$-ln(p_g(d)) = \sum_{s} p_g(s|d) E_g(s;d) - H_g(s|d) = \langle E_\mathbf{g}(\mathbf{s};d) \rangle_\mathbf{g} - H_\mathbf{g}(\mathbf{s}|d) = F_g(d) \tag{4.41}$$

Thus, a network's free energy can be expressed as either (a) the negative log of the probability that the network generates output d, or (b) the average energy minus the entropy of network states induced by d. Via equation 4.27, another defintion of $F_g(d)$ is the average surprisal minus entropy.

Since a well-trained network should generate d with a high probability, the goal of adaptation is to increase $ln(p_g(d))$ (toward 0) and thus to decrease its negative, the free energy. This general goal of minimizing free energy pertains to a variety of machine-learning techniques (Mackay 2003), not only neural networks.

The presence of both energy and entropy in free energy has important significance and indicates a direct parallel to the concept of mutual information, defined as

$$I(S, D) = H(S) + H(D) - H(S, D) \tag{4.42}$$

where H(S,D) is the entropy of the coupled system and environment.

A common goal in an information-processing system is to maximize the mutual information between two communicating components. This occurs when each component can inhabit many different states with approximately equal probability (i.e., high entropy of the individual components), but the coupled situation displays a highly skewed distribution, due to the constraints that the components impose on one another's behavior. As a simple example, two people have a high rate of information exchange when (a) they each command a large vocabulary (and are not too biased toward using particular words over others), but (b) when one person speaks, it greatly restricts the set of words that the listener (believes she) hears, thus reducing the entropy of the coupled speaker-listener system.

When free energy serves as an objective to minimize in a neural network, the first subgoal, reducing average energy, corresponds to reducing H(S,D): both involve a reduction of conflict by moving into states that satisfy the constraints imposed by (a) each $d \in D$, and (b) the network's weights; and the more constraints, the greater potential for lowering H(S,D). The second subgoal, increasing entropy, mirrors that of elevating H(S): a system with a wide, relatively even, distribution of internal states has more flexibility to conform to its environment. Since free-energy metrics formalize this level of conformity but ignore the system's potential to change the environment, H(D) is not a factor that the network can influence. Figure 4.8 illustrates these relationships.

As an intermediate summary of the use of free energy in neural networks, begin with the goal of tuning a network so that it generates a probability distribution $p_g(D)$ similar to the environmental distribution, p(D). The probabilities in p_g stem directly from energy, as

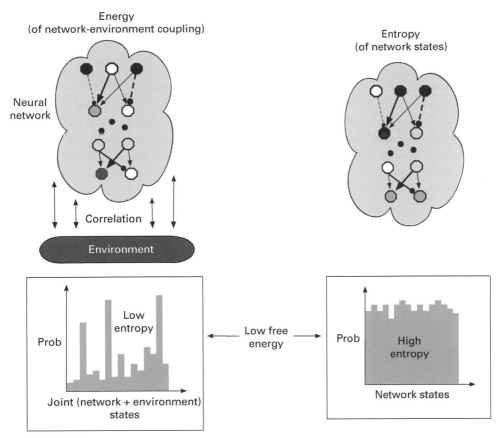

Figure 4.8
Dual aspects of minimizing free energy. (Left) Reducing average energy entails reducing conflict between an environmental state (d) and internal explanatory states, thus biasing the distribution of these explanations to conform to d and the network's weights, and thereby reducing the entropy of the system-environment coupling. (Right) High entropy of the explanatory states visited under the influence of d, but where the entropy calculation does not include d's bits as part of any state. This combination is similar to a situation of high mutual information between system states and the environment, where both the environment and system should have high entropies, while their coupled state does not.

defined by the Boltzmann equation, and they are inversely proportional to the information theoretic notion of surprisal. The process of moving $p_g(D)$ toward p(D) entails minimizing the Kullback-Leibler divergence between the two distributions, and this turns out to be equivalent to minimizing the negative log likelihood of the data, which is equal to the free energy. Thus, training the neural network involves minimizing free energy, which can also be expressed as minimizing surprisal and maximizing entropy.

4.5.1 Variational Free Energy

In statistical mechanics, Z, from the Boltzmann equation, and thus free energy (F) are based on the *true* probability distribution of states (s), denoted here as $p_g(s)$. However, in many situations, only an estimate of or proxy for the true distribution exists: $p_r(s)$. Under these circumstances, free energy incorporates a mixture of the two distributions and becomes *variational free energy*: F_g^r, defined as the combination of the free energy and the divergence between the two distributions:

$$F_g^r = -lnZ + D_{KL}(p_r \| p_g) = F + D_{KL}(p_r \| p_g) \tag{4.43}$$

Since $D_{KL}(p_r \| p_g) \geq 0$ by the Gibbs inequality, the variational free energy serves as an upper bound on the free energy, and any operations that reduce F_g^r should help to decrease F as well, depending on any accompanying changes to the divergence. Practically speaking, any algorithm that seeks to reduce free energy in a system but has poor information about Z can, at least, employ a more accessible distribution and try to decrease the resulting variational free energy and/or the KL divergence between the two distributions. The mathematical relationship between these quantities are expressed in the accompanying box below.

Relationship between Free Energy and Variational Free Energy

Just as standard free energy equals average energy minus entropy, the variational free energy has a similar equivalence, but one based on both distributions. The following derivation shows this relationship, starting with the previous definition of variational free energy:

$$F_g^r = D_{KL}(p_r \| p_g) - lnZ = \sum_s p_r(s) ln \left(\frac{p_r(s)}{p_g(s)} \right) - lnZ \tag{4.44}$$

Since the log of a quotient is the difference of logs, the first summation can be decomposed into two, and since $\sum_s p_r(s) = 1$ and Z is independent of $p_r(s)$, a third summation can be introduced, yielding

$$F_g^r = \sum_s p_r(s) ln(p_r(s)) - \sum_s p_r(s) ln(p_g(s)) - \sum_s p_r(s) lnZ \tag{4.45}$$

Recognizing the first term as the negative entropy and rearranging:

$$F_g^r = - \sum_s p_r(s)(ln(p_g(s)) + lnZ) - H_r(s) \tag{4.46}$$

Noting that the sum of logs is the log of the product, and then using $E_g(s) = -ln[p_g(s)Z]$ from earlier (equation 4.8):

$$F_g^r = \sum_s p_r(s)[-ln[p_g(s)Z]] - H_r(s) = \sum_s p_r(s) E_g(s) - H_r(s) \tag{4.47}$$

Changing to the bracket notation for the expected value with respect to r yields the desired final form:

$$F_g^r = \langle E_g(s) \rangle_r - H_r(s) \tag{4.48}$$

This is a mixture of the probability distributions in the sense that $E_g(s)$ corresponds to $p_g(s)$ via the Boltzmann equation, but $p_r(s)$ is the distribution over which both the energy and entropy terms are averaged.

Variational free energy plays an important role in neural networks that have separate recognition (r) and generation (g) phases, with specific synaptic weights for each phase, and thus different probability distributions, p_g and p_s, over the internal states created during each phase. Each weight set creates different relationships between local activation patterns, although, ideally, those distributions converge with repeated experience interpreting and predicting the environmental experiences, D.

For any $d \in D$, and all possible explanatory network states (S), Kullback-Leibler divergence of the distributions over S provides a metric for assessing convergence:

$$D_{KL}(p_r(S|d), p_g(S|d)) = \sum_{s \in S} p_r(s|d) ln\left(\frac{p_r(s|d)}{p_g(s|d)}\right) = \sum_{s \in S} p_r(s|d) ln(p_r(s|d))$$

$$- \sum_{s \in S} p_r(s|d) ln(p_g(s|d)) \tag{4.49}$$

Using the definition of entropy (H) and $p_g(s|d) = \frac{p_g(s,d)}{p_g(d)}$:

$$= -H_r(S|d) - \sum_{s \in S} p_r(s|d) ln\left(\frac{p_g(s,d)}{p_g(d)}\right) = -H_r(S|d) - \sum_{s \in S} p_r(s|d) ln(p_g(s,d))$$

$$+ \sum_{s \in S} p_r(s|d) ln(p_g(d)) \tag{4.50}$$

Since $\sum_{s \in S} p_r(s|d) = 1$ and $p_g(d)$ is independent of $p_r(s|d)$:

$$= -H_r(S|d) - \sum_{s \in S} p_r(s|d) ln(p_g(s,d)) + ln(p_g(d)) \tag{4.51}$$

Since $-ln(p_g(d)) = F_g(d)$ and $ln(p_g(s,d)) = -E_g(s;d)$:

$$= -H_r(S|d) + \sum_{s \in S} p_r(s|d) E_g(s;d) - F_g(d) = -H_r(S|d) + \langle E_g(s;d)\rangle_r - F_g(d) \tag{4.52}$$

Thus:

$$D_{KL}(p_r(S|d), p_g(S|d)) = \underbrace{-H_r(S|d) + \langle E_g(s;d)\rangle_r}_{F_g^r(d)} - F_g(d) \tag{4.53}$$

So once again, the variational free energy provides an upper bound on the free energy (since KL divergence never dips under zero):

$$F_g^r(d) = D_{KL}(p_r(S|d), p_g(S|d)) + F_g(d) \tag{4.54}$$

Equation 4.54 provides a useful framework for training and analyzing bidirectional networks, whose typical goal is to generate patterns consistent with those that it learns to recognize: the explanatory state invoked by an experience, d, should also be the state used to predict (i.e., generate) d. Thus, through experience and learning, $D_{KL}(p_r(S|d), p_g(S|d))$ should decline, as should the generative error: $-ln(p_g(d)) = F_g(d)$, the free energy. So, by equation 4.54, learning should reduce variational free energy, $F_g^r(d)$ for any $d \in D$.

Minimizing $F_g^r(d)$ is therefore a practical objective for training these networks, and the natural question then becomes, How do the weights affect $F_g^r(d)$, that is, what is $\frac{\partial F_g^r(d)}{\partial w}$? Fortunately, the operationalization of this derivative often leads to another version of contrastive Hebbian learning (CHL), as shown in section 4.6 for the Helmholtz machine.

4.6   The Helmholtz Machine

Invented in the mid-1990s by Peter Dayan and several other members of Geoffrey Hinton's Toronto group, the Helmholtz machine (Hinton et al. 1995; Dayan et al. 1995) applies Boltzmann machine principles to multilayered bidirectional networks that alternate between recognition and generation phases. This models the human perceptual system as a hierarchical statistical inference mechanism that combines bottom-up and top-down processing. Figure 4.9 displays the basic topology.

The Helmholtz machine was designed as a neural network implementation of the expectation maximization (EM) algorithm, an unsupervised learning method in which alternating phases of data classification and class modification gradually refine both the class labels of each data point and the parameters defining each class. EM elegantly implements the general philosophy that deep understanding stems from an interleaving of data interpretation and generation, and this process obviates the need for supervisory feedback from the environment.

In the Helmholtz machine, the central classes are the causal explanatory states manifest in internal activation patterns. These causes are operationally defined in terms of the data that they produce via the generative weights, so modifying the definition of class involves changing the generative weights. Hence, the maximization phase of EM corresponds to the Helmholtz wake phase, in which concept definitions change. Conversely, the

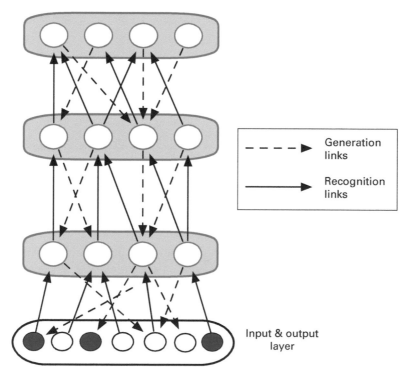

Figure 4.9
Sketch of a Helmholtz machine, with recognition and generative weights linking each pair of adjacent layers. There are typically no intralayer connections in these networks.

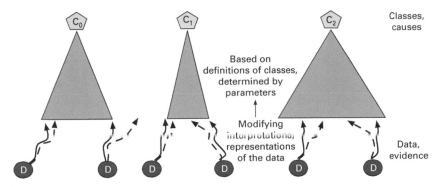

Figure 4.10
The expectation phase of the expectation maximization (EM) algorithm in which the parameters for *interpreting* the data (twisted arrows) are modified. In a typical (unsupervised) clustering algorithm, this phase updates the current class of each data item. In the Helmholtz machine, this (sleep) phase changes the interpretation of each data point by adjusting the recognition weights.

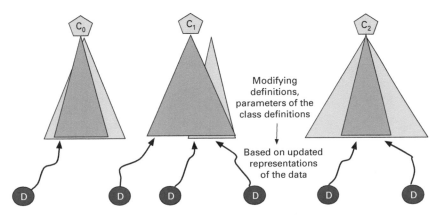

Figure 4.11
The maximization phase of the expectation maximization (EM) algorithm, in which the parameters for *generating* data (triangles) are modified. In a typical (unsupervised) clustering algorithm, this phase updates the parameters (e.g., mean and variance) that define each class, based on the data points currently grouped into that class. In the Helmholtz machine, this (wake) phase alters the network's top-down, generative weights.

expectation phase of EM entails changes to the classification of existing data points, and such changes mirror the updates of recognition weights in the sleep phase. Figures 4.10 and 4.11 summarize these relationships.

The introduction of bidirectional weights cleanly separates parametric responsibility for the recognition and generation phases, thus permitting one phase to produce *targets* for training the parameters of the opposite phase, which essentially performs prediction. Thus, in these networks, prediction involves both bottom-up and top-down propagation, depending on the phase. As shown below, Helmholtz operation involves two probability distributions over internal states; training the network then entails lowering the variational free energy of these states.

Figure 4.12 portrays the recognition/wake phase in which the loop of activation begins and ends at the lowest, sensory level. During the upward pass through the hierarchy, each

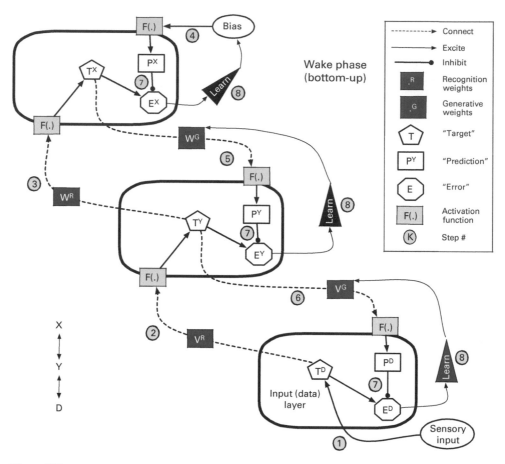

Figure 4.12
The bottom-up, wake phase of Helmholtz machine operation in which upward signaling produces target values in each level. The following downward return signals then produce predictions for each target, with the difference (target minus prediction) yielding an error term used in the local Hebbian updating of the top-down (generative) weights.

layer updates its target values before propagating them further across a set of recognition weights. When the signal reaches the highest level, it descends back through the hierarchy, but this time using the generative weights. These descending signals constitute *predictions* for each layer. As shown in more detail in figure 4.14 (left), the difference between targets and predictions yields an error vector, which combines with the activations of the layer above to form the local Hebbian learning rule of equation 4.55 (with symbols defined in figure 4.12).

$$\triangle W^G = \lambda T^X [T^Y - \underbrace{F(T^X \bullet W^G)}_{prediction}] \tag{4.55}$$

In the dream phase of figure 4.13, the activation loop begins at the upper level, where a *dreamed* pattern propagates downward via the generative weights, depositing a target value at each level. This produces an imagined input (i.e., a sensory prediction) at the lowest layer, which then propagates upward, producing predictions in each layer. Once again, the

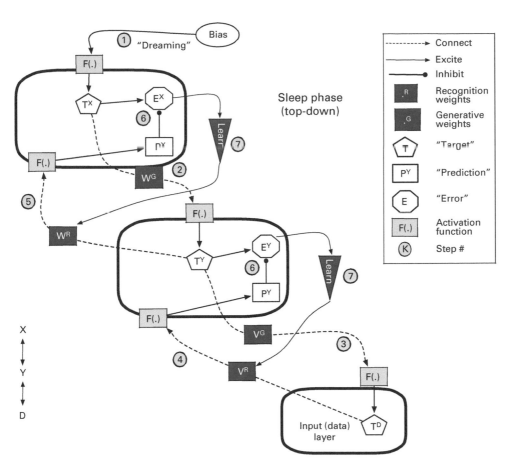

Figure 4.13
The top-down, sleep phase of Helmholtz machine operation in which randomly generated patterns at the top level initiate downward signaling that produces target values in each level. The following upward return signal (based on generated, *dreamed* sensory input) creates predictions for each target. The error (target minus prediction) then governs local Hebbian changes to the bottom-up (recognition) weights.

difference between a layer's targets and predictions creates an error vector, which, in this case, combines with the activations of the layer below as the basis for Hebbian learning, as detailed in equation 4.56 (with symbols defined in figure 4.13).

$$\triangle V^R = \lambda T^D [T^Y - \underbrace{F(T^D \bullet V^R)}_{prediction}] \tag{4.56}$$

Notice that the learning rules in equations 4.55 and 4.56 can be interpreted either as (a) versions of the classic Delta rule, which multiplies the presynaptic activation by the postsynaptic error, or (b) examples of contrastive Hebbian learning, with a Hebbian term involving the targets of adjacent layers and an anti-Hebbian term based on a prediction. Regardless, the important characteristics of these rules are their locality and their combination of targets and predictions. The uniqueness of the Helmholtz machine lies in the full inverse relationship between the two phases such that *reality* and the targets it produces can come from deep within the network just as well as from the sensory surround.

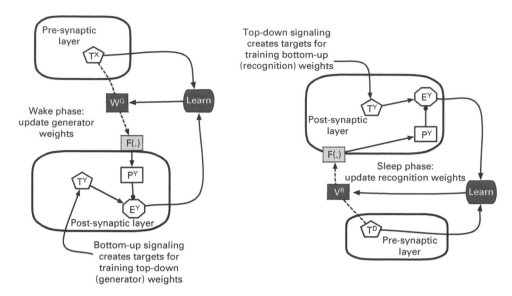

Figure 4.14
Learning in the wake (left) and sleep (right) phases of the Helmholtz machine.

The real beauty of the Helmholtz machine stems from the ability of its local learning rules to minimize a complex global objective function: the variational free energy. Understanding this key relationship (an interesting example of emergence) requires a few more details of activation propagation in these networks, along with a little mathematics (provided in the accompanying box below).

Although never a competitor to standard connectionist backpropagation networks, the Helmholtz machine provides a hierarchical model of recognition, prediction, and learning that nicely matches neuroscientific evidence of the brain's layered architecture with bidirectional signal flow. It embodies the old adage of learning by synthesis, not only by interpretation. By linking bottom-up recognition and top-down generation, it allows each phase to produce targets for the other, thereby reducing the amount of sensory sampling required to produce an internal understanding / explanation of the environment.

Deriving Local Learning Rules for Helmholtz Machines

As shown in figure 4.15, each layer's targets are formed by Gibbs sampling of the activation vectors produced by upward signal flow (in the wake phase) capped off with a sigmoid activation function. Thus, with sigmoid output s_j, the probability of target unit j having its current binary value x_j is

$$p(x_j) = s_j^{x_j}(1 - s_j)^{(1-x_j)} \qquad (4.57)$$

Since the target units are conditionally independent of one another (given the layer's pre-synaptic values), the probability of the complete target-vector state is

$$p(x) = \prod_i s_i^{x_i}(1 - s_i)^{(1-x_i)} \qquad (4.58)$$

It is also important to note that during the downward flow of the wake phase, no Gibbs sampling occurs. Hence, these postsynaptic values (i.e., the predictions) remain as real-valued outputs of the sigmoid function. Hence, a layer's error is the difference between a binary target vector and a real-valued prediction vector. This same relationship holds in the sleep phase, except that the targets and predictions come from above and below, respectively.

During the recognition phase, the generational weights are being updated and thus the generational distribution of states (p_g) takes on the role of the goal distribution, while the recognition distribution (p_r) is the approximating distribution. The goal during recognition is to decrease the variational free energy from p_r to p_g: F_g^r. This requires calculating the derivative of variational free energy with respect to each *generative* weight, w_{kj}:

$$\frac{\partial F_g^r(d)}{\partial w_{kj}} = \frac{\partial \langle E_g(s;d) \rangle_r}{\partial w_{kj}} - \underbrace{\frac{\partial H_r(S|d)}{\partial w_{kj}}}_{0} \tag{4.59}$$

Since the entropy of states due to the recognition distribution is independent of the generative weights, the second term of equation 4.59 vanishes. Then, $E_g(s;d) = -ln(p_g(s,d))$ leads to:

$$\frac{\partial F_g^r(d)}{\partial w_{kj}} = -\frac{\partial \langle ln(p_g(s,d)) \rangle_r}{\partial w_{kj}} \tag{4.60}$$

Assuming that weight updates occur after the processing of each data case $d \in D$, the minimization problem decomposes into computing and applying the gradient (with learning rate λ) for each weight on each case (d):

$$\triangle w_{kj} = -\lambda \frac{\partial E_g(s;d)}{\partial w_{kj}} = -\lambda \frac{-\partial ln(p_g(s,d))}{\partial w_{kj}} = \lambda \frac{\partial ln(p_g(s,d))}{\partial w_{kj}} \tag{4.61}$$

Since d is independent of the generative weights, when considering the derivative of $ln(p_g(s,d))$ with respect to w_{kj}, d can be ignored, thus shifting the focus to $p_g(s)$. Assuming that x represents the binary state of the entire network, whereas s denotes the real-valued activation levels in the network, the probability of x is

$$p_g(x) = \prod_i s_i^{x_i}(1-s_i)^{(1-x_i)} \tag{4.62}$$

Calculating its natural log:

$$ln(p_g(x)) = \sum_i x_i ln(s_i) + (1-x_i)ln(1-s_i) \tag{4.63}$$

Weight w_{kj}, on the synapse from neuron k to neuron j, only affects the jth neuron, so the derivative simplifies to

$$\frac{\partial ln(p_g(x))}{\partial w_{kj}} = \frac{\partial [x_j ln(s_j) + (1-x_j)ln(1-s_j)]}{\partial w_{kj}} \tag{4.64}$$

Remember that x_j stems from Gibbs sampling of s_j, whose value *was* derived from *recognition* weights, though s_j (but not x_j) was modified on the downward pass. Hence, x_j is independent of the *generative* weight, w_{kj}, which further simplifies the derivative:

$$\frac{\partial ln(p_g(x))}{\partial w_{kj}} = x_j \frac{\partial ln(s_j)}{\partial w_{kj}} + (1-x_j)\frac{\partial ln(1-s_j)}{\partial w_{kj}} \tag{4.65}$$

Since $\frac{\partial ln(a)}{\partial b} = \frac{\frac{\partial a}{\partial b}}{a}$:

$$\frac{\partial ln(p_g(x))}{\partial w_{kj}} = x_j \frac{\frac{\partial s_j}{\partial w_{kj}}}{s_j} + (1 - x_j) \frac{\frac{\partial (1 - s_j)}{\partial w_{kj}}}{(1 - s_j)} \tag{4.66}$$

As commonly shown in derivations of the backpropagation algorithm via repeated uses of the chain rule of calculus, $\frac{\partial s_j}{\partial w_{kj}}$ is the product of two terms: (a) the effect of w_{kj} on the sum of weighted inputs to neuron j: x_k, and (b) the derivative of the activation function (at unit j) with respect to that sum. Remember that s_j comes from a sigmoid activation function, whose derivative is the product of its output and 1 minus its output. Thus:

$$\frac{\partial s_j}{\partial w_{kj}} = \underbrace{x_k s_j (1 - s_j)}_{\Phi} \tag{4.67}$$

Using the substitution of $\Phi = x_k s_j (1 - s_j)$:

$$\frac{\partial ln(p_g(x))}{\partial w_{kj}} = x_j \frac{\Phi}{s_j} + (1 - x_j) \frac{-\Phi}{(1 - s_j)} = \frac{x_j \Phi (1 - s_j) - s_j \Phi (1 - x_j)}{\underbrace{s_j (1 - s_j)}_{\frac{\Phi}{x_k}}} \tag{4.68}$$

Simplifying:

$$\frac{\partial ln(p_g(x))}{\partial w_{kj}} = x_k [x_j (1 - s_j) - s_j + s_j x_j] = x_k [x_j - s_j] \tag{4.69}$$

Hence, the weight update is simply the Delta rule:

$$\triangle w_{kj} = \lambda x_k [x_j - s_j] \tag{4.70}$$

that is, learning-rate • presynaptic-output • [postsynaptic-target − postsynaptic-output].
    Similar calculations (involving a few more assumptions) also yield a Delta rule for the sleep phase. Consequently, all of the adaptivity of the Helmholtz machine can be attributed to local, Hebbian weight changes that reduce the variational free energy of the entire system.

4.7   The Free Energy Principle

Karl Friston's free energy principle (FEP) (Friston 2005; Friston, Kilner, and Harrison 2006; Friston 2010) provides the mathematical groundwork for many promising theories of the brain, particularly those revolving around prediction. As depicted in figure 4.16, Friston leverages free energy to explain perception and action, with each based on a different reformulation of the variational free energy equation shown earlier (see the box "The Free Energy Principle: Three Equivalent Expressions" for mathematical details).
    Equation 4.76 indicates how decreasing the variational free energy in a neural system entails reducing the surprise (of the data) and/or reducing the divergence between the recognition and generative distributions over the internal system states. Note that surprise

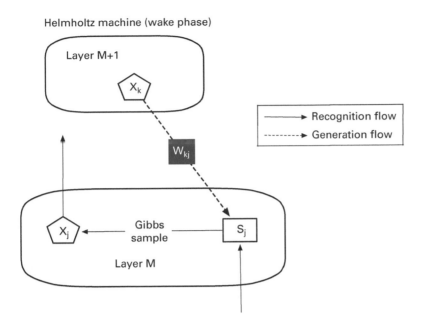

Helmholtz machine (wake phase)

Figure 4.15
Gibbs sampling during upward signal flow of the Helmholtz machine's wake phase. The values in vector S_j are the outputs of a sigmoid activation function and thus represent probabilities, which govern the Gibbs sampling that produces binary values for the vector X_j.

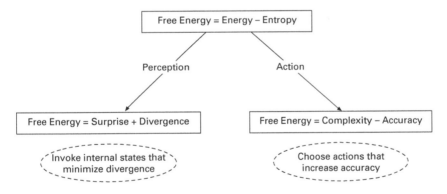

Figure 4.16
Overview of Friston's free energy principle (FEP), which involves three equivalent formulations of free energy (boxes). Perception and action operate through different terms (divergence and accuracy, respectively) of these expressions, as detailed in the text.

is minimal (close to zero) when the network accurately predicts the environmental data $d \in D$.

Friston also employs the formulation in equation 4.76 to reconcile perception with the minimization of variational free energy. That explanation focuses on running the system in recognition mode given a sensory input, d, and with the goal of producing a distribution of internal explanatory states that best matches the target distribution of d's causes, which Friston equates with the generative distribution $p_g(s|d)$.[3]

The Free Energy Principle: Three Equivalent Expressions

To get a quantitative understanding of the FEP, begin with the following expression for the variational free energy of a neural system:

$$F_g^r(d) = \langle E_g(s; d) \rangle_r - H_r(s|\Theta) = \langle -ln[p_g(s,d)] \rangle_r + \langle ln[p_r(s|\Theta)] \rangle_r \tag{4.71}$$

where Θ represents system parameters such as synaptic strengths, neuromodulatory levels, and so on. Since $p(x,y) = p(x|y)p(y)$:

$$F_g^r(d) = \langle -ln[p_g(s,d)] \rangle_r + \langle ln[p_r(s|\Theta)] \rangle_r = \langle -ln[p_g(s|d)p_g(d)] \rangle_r + \langle ln[p_r(s|\Theta)] \rangle_r \tag{4.72}$$

Because (a) the logarithm of a product is the sum of logarithms, and (b) the difference of logarithms is the logarithm of their quotient:

$$= \langle -ln[p_g(d)] \rangle_r + \langle -ln[p_g(s|d)] \rangle_r + \langle ln[p_r(s|\Theta)] \rangle_r = \langle -ln[p_g(d)] \rangle_r + \langle ln\frac{p_r(s|\Theta)}{p_g(s|d)} \rangle_r \tag{4.73}$$

Rewriting based on the semantics of $\langle \ldots \rangle_r$:

$$= -\sum_s p_r(s|\Theta)ln[p_g(d)] + \sum_s p_r(s|\Theta)ln\frac{p_r(s|\Theta)}{p_g(s|d)} \tag{4.74}$$

Knowing that $ln[p_g(d)]$ is independent of the distribution $p_r(s|\Theta)$, and using the definition of D_{KL}:

$$= -ln[p_g(d)] \underbrace{\sum_s p_r(s|\Theta)}_{=1} + D_{KL}(p_r(s|\Theta), p_g(s|d)) \tag{4.75}$$

Thus:

$$F_g^r(d) = \underbrace{-ln[p_g(d)]}_{surprise} + \underbrace{D_{KL}(p_r(s|\Theta), p_g(s|d))}_{divergence} \tag{4.76}$$

In addition, Friston redefines $F_g^r(d)$ in terms of complexity and accuracy. Similar to the derivation above, this one begins with equation 4.71 and rewrites the joint probability, but this time using the opposite conditional probability: $p(x,y) = p(y|x)p(x)$. Thus:

$$F_g^r(d) = \langle -ln[p_g(s,d)] \rangle_r + \langle ln[p_r(s|\Theta)] \rangle_r = \langle -ln[p_g(d|s)p_g(s)] \rangle_r + \langle ln[p_r(s|\Theta)] \rangle_r \tag{4.77}$$

Again, using properties of logarithms of products and quotients along with a little rearrangement:

$$= \langle -ln[p_g(d|s)] \rangle_r + \langle -ln[p_g(s)] \rangle_r + \langle ln[p_r(s|\Theta)] \rangle_r = \langle -ln[p_g(d|s)] \rangle_r + \langle ln\frac{p_r(s|\Theta)}{p_g(s)} \rangle_r \tag{4.78}$$

Rewriting based on the semantics of $\langle \ldots \rangle_r$:

$$= -\sum_s p_r(s|\Theta)ln[p_g(d|s)] + \sum_s p_r(s|\Theta)ln\frac{p_r(s|\Theta)}{p_g(s)} \tag{4.79}$$

Rearranging and using the definition of D_{KL} yields the desired form:

$$F_g^r(d) = \underbrace{D_{KL}(p_r(s|\Theta), p_g(s))}_{complexity} - \underbrace{\sum_s p_r(s|\Theta) ln[p_g(d|s)]}_{(predictive)accuracy} \qquad (4.80)$$

In this context, complexity is the divergence between the prior distribution of explanatory states, $p_g(s)$, and the posterior (i.e., after performing recognition) distribution, $p_r(s|\Theta)$. Accuracy denotes the ability of the high-probability internal states to predict the actual data: those internal states that the system often exhibits tend to cause the environmental interface to *expect* the sensory patterns $d \in D$.

In equation 4.76, the key symbol is Θ, which represents modifiable parameters that affect the recognition process. Friston views perception as properly choosing Θ in order to reduce the divergence term of equation 4.76, and thereby reduce $F_g^r(d)$. This mirrors the Bayesian brain hypothesis, which defines perception as a reduction in the divergence of the recognition distribution from a distribution based on the generative model.

To understand action in terms of minimizing variational free energy, use equation 4.80, but with d transformed into $d(\alpha)$: the sensory input becomes a function of the chosen action, α:

$$F_g^r(d) = \underbrace{D_{KL}(p_r(s|\Theta), p_g(s))}_{complexity} - \underbrace{\sum_s p_r(s|\Theta) ln[p_g(\mathbf{d}(\alpha)|s)}_{(predictive)accuracy} \qquad (4.81)$$

Equation 4.81 provides a framework for a variant of *active perception* that Friston calls *active inference* (Friston 2010), where the goal of action is to improve predictive accuracy. In the FEP, this improvement entails choosing actions that will produce sensory inputs that are predicted by the currently, highly probable, causal hypotheses. In the notation of equation 4.81, predictive improvement means choosing α to produce sensory data $d(\alpha)$ such that, for any causal state ($s*$) that is a good explanation of $d(\alpha)$ (i.e., where $p_g(d(\alpha)|s*)$ is high), it is also the case that $p_r(s*|\Theta)$ is high. Hence, action reduces variational free energy by increasing predictive accuracy.

In Friston's interpretation, good actions are those that confirm the system's favored hypotheses. This differs from other views of active perception that involve decreasing the average uncertainty (i.e., entropy) of the hypothesis pool. This latter view suggests a *fair and unbiased* aspect to action in terms of making the best effort to find the best explanation for the data, that is, the ideal scenario for scientific hypothesis testing. Conversely, Friston's theory clearly embraces subjectivity: agents act to support their current causal hypotheses of the world. This makes perfect sense as a cognitive model of action, since living organisms are not objective, unbiased automata, but rather, survival machines that will do just about anything to improve their advantage in a world that they have limited (but useful) capacities to predict and control.

As Andy Clark (2016) emphasizes, brains are designed to optimize the organism's engagement with the environment; the agent *maximally exploits* the world. Thus, the internal predispositions become central, and although they can change, the well-adapted organism needs to do so only sparingly. For the most part, its implicit and explicit predictions about the world come true. Throughout both the evolutionary history of its species and its own lifetime of development and learning, the probabilities of expectations matching reality tend to increase: accuracy rises. Furthermore, the deviations between the priors and posteriors of causal hypotheses diminish: complexity decreases. Thus, as an agent (and species) adapts, the friction between it and the world drops, yielding a seamless engagement quantified as a *low free-energy existence*.[4]

4.8   Getting a Grip

As a PhD student—earning a livable income, unfettered by deadlines, and having only a vague goal of someday producing a large document full of deep insights—I had a little too much time to philosophize. Possibly the most significant result of that utopian existence was a core tenet of my general view of life: "You only achieve a significant *grip* on life when you realize that you will never have a full grip on anything. It's all about adapting, continuously, and embracing it."

There is nothing unique in that memo; it just took me twenty-eight years to get it.

This *grip*, another word that Clark uses extensively, seems to reflect successful bidirectional interaction with the world, in particular, with the sensory flow from that world to the body and brain. And it, too, can never be fully optimized: the adaptive changes are the only constant. The perpetual flux of our environments (and growth and deterioration of our bodies) demand continuous exploration and internal adjustments to maintain a reasonable level of control over our individual mind-body-world couplings. In some cases, this entails improving our models of the world so as to better predict future states; while in others, the biases (i.e., probability distributions) regarding various action sequences may change to improve our peaceful coexistence with the world, for example, by avoiding situations (such as casino gambling) where we sense a limited ability to ever get a foolproof internal model and winning grip.

The neural networks of this chapter are often viewed as the dinosaurs of connectionism: old classics that served as key predecessors to the high-powered, human-competitive models of the twenty-first century. Boltzmann machines achieved multilayered learning prior to the invention of backpropagation, and successful RBMs pre-dated the CNNs and LSTMs that rocketed deep learning to prominence. However, it would be a major mistake to relegate these architectures to the junk heap of computational history as merely obsolete scaffolding for modern DL. Instead, they operationalize fundamental principles for achieving that everelusive grip.

In discussing systems that minimize free energy and prediction error via a seamless integration of perception and action, Clark (2016, 297) concludes the following:

By self-organizing around prediction error, and by learning a generative, rather than a merely discriminative (i.e., pattern classifying) model, these approaches realize many of the dreams of previous work in artificial neural networks, robotics, dynamical systems theory, and classical cognitive science. They perform unsupervised learning using a multilevel architecture and acquire a satisfying *grip* [my emphasis]—courtesy of the problem decompositions enabled by their hierarchical form—upon structural relations within a domain. They do so, moreover, in ways that remain firmly grounded in the patterns of sensorimotor experience.

This emergent parallel bootstrapping of predictive and interpretive competence forms a crucial link between the sensorimotoric and the conceptual, a connection that underlies true understanding. By predicting, our brains establish goals (set points) for regulatory systems whose actions (including acts of sensing) are strongly biased toward confirming those expectations, not toward attaining some objective truth about the world. Only when the egotism of the generative mode proves deleterious to the agent (by spurring inappropriate actions) does it need to adjust its predictions to better match reality and modify its interpretations of reality (including the attention levels it allocates to different prediction errors) to better highlight mismatches and ultimately update its world model and action strategies. But prior to these (minor or major) breakdowns, the agent can let its own intentions and beliefs run the show, under the practical illusion that it has *the full grip*, when it actually only has a sufficient grip for its current situation. And, as explained in chapter 5, the modular, hierarchical structure of these predictive engines provides cross-generational *grip maintenance*, since evolvability of such systems is greatly enhanced by modular decomposition.

5 Predictive Coding

5.1 Information Theory and Perception

In the mid 1980s, I enrolled in a graduate seminar, Perceptual Cognition, at the University of Oregon. Our weekly meetings were classic grad-school fare, with a dozen students, faculty members, and semi-mysterious others lounging on deep couches, drinking tea, and discussing topics well beyond the safe confines of my primary field of doctoral study: computer science. Luminaries such as Michael Posner occasionally dropped in to lend full legitimacy to the gatherings, but some of the other older participants seemed out of place. Eugene, Oregon, has always had its share of interesting and interested individuals who attend classes for no other reason than their immense curiosity, so I quickly grouped a few of the others into that category, including a long, lanky man with a thin goatee and a distinct southern accent who spoke rarely but got the full attention of Posner and others when he did. As has happened several times in my career, not until years later did I realize that I had been in the presence of a true giant of science: the bearded gentleman with the deep voice and deeper insights was Fred Attneave, a pioneer at the crossroads of psychology and information theory.

In 1954, Attneave introduced his *theory of redundancy reduction for visual perception* with a seminal paper (Attneave 1954) outlining numerous properties of images that facilitate data compression, thus easing the burden of signal transmission in perception. Essentially, any instances of temporal or spatial correlation among sensory inputs provide fodder for systematic reductions in signal load:

When we begin to consider perception as an information-handling process, it quickly becomes clear that much of the information received by any higher organism is *redundant*. Sensory events are highly interdependent in both space and time: if we know at a given moment the states of a limited number of receptors (i.e., whether they are firing or not firing), we can make better-than-chance inferences with respect to the prior and subsequent states of these receptors, and also with respect to the present, prior and subsequent states of other receptors. (Attneave 1954, 183)

Attneave's work introduced predictive coding to cognitive science, though the term itself would not appear for several more decades. Another precursor was Oliver's (1952) *efficient coding*, which found widespread usage in telecommunications starting in the early 1950s.

The general process of predictive coding is to recode (compress) data by removing all aspects of it that are predictable. Thus, the efficiently encoded version is the original minus

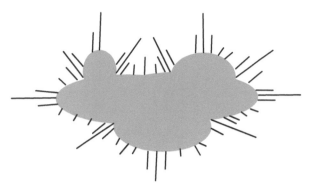

Figure 5.1
A general sketch reflecting the basic results of Attneave's classic experiment in which subjects placed ten dots at locations along the object's contour that they believed would most accurately delineate its shape (Attneave 1954). Lengths of emanating lines denote the number of subjects choosing that particular location.

the prediction. This difference, termed the *prediction error*, is the remainder of the original signal that gets propagated further in the system.

As a simple example of this basic concept, imagine the task of transmitting a series of integers (1000, 1002, 1003, 998, and 997) along a channel that carries only binary signals. Each integer would require 10 bits, for a total of 50 bits, if transmitted independently of the others. But by noticing the correlations among the five values—they are all very similar in magnitude—a more efficient strategy becomes obvious: transmit a base value (e.g., an average) plus the deviations of each from that base. Thus, the six values to transmit are 1000 (the average, and also the prediction) along with $0, 2, 3, -2, -3$. Assuming that the sign associated with each deviation requires one extra bit, each of these latter five values uses no more than 3 bits; and the entire sequence compresses to $10 + 5 \times 3 = 25$ bits. Essentially, the deviations from predictions / expectations / averages demand the most transmission effort.

As sketched in figure 5.1, Attneave illustrated the primality of deviations from the norm via an experiment in which subjects were asked to place ten dots at any locations along a shape's outline that they believed would most accurately approximate the pattern (when viewed in isolation). The results clearly showed the perceptual saliency of contour changes (i.e., sharp bends in the outline) in defining the figure. As Attneave suggests, a compressed description of the image would exploit strong spatial correlations to compress homogeneous regions (or stretches of the outline) into simple descriptors that combine the precise locations of a few points along with the neighborhood relationships between those points. With greater correlation comes greater possibilities for compression.

In 1982, Srinivasan and colleagues coined the term *predictive coding* in showing how inhibitory interneurons in the retina realize the data compression predicted by Attneave's theory (Srinivasan, Laughlin, and Dubs 1982). One primary motivation for their work— and possibly a key constraint in the evolution of the retina—is the uncertainty inherent in neural signalling, known as *intrinsic noise*. Even when sensory receptors provide accurate readings, transmission across multiple synapses inevitably degrades information. What can neural networks do to combat this?

To understand how predictive coding ameliorates this problem, imagine an internal neuron with a range of 0–200 Hz, where spike frequency conveys the main information. In theory, this neuron can transmit 200 uniquely identifiable signals per second. Now assume that the operative range of the sensory input is, for example, 0 to 1000 units. Hence, the neural signal achieves a resolution of $\frac{1000}{200} = 5$ sensory units: each one-second interval of spikes transmits a sensory value with a precision of 5 units. But what happens when intrinsic noise hampers internal transmission?

Suppose that the intrinsic noise equals plus or minus 5 Hz: a signal that should be 87 spikes/sec could be anywhere from 82 to 92 spikes/sec. This represents a noise range of 10 Hz and restricts the one-second neural signal to $\frac{200}{10} = 20$ uniquely identifiable signals, since, for example, an 87 and a 90 are no longer guaranteed to convey different information. Any downstream neuron receiving those spikes cannot safely react to 87 differently than 90, since they may be transmitting identical sensory information. In effect, this reduces the functional resolution of the sensory inputs to $\frac{1000}{20} = 50$ units, that is, a vague blur compared to the noise-free condition.

Since both intrinsic noise and the upper bound on firing rates (in the range of 250–1000 Hz) are inevitable and immutable consequences of neurophysiology, the task of increasing functional sensory resolution becomes the responsibility of earlier processes in the signaling pipeline. In short, if the effective range of sensory signals can be reduced, then the weak signal-to-noise ratio of intrinsic signaling can still achieve respectable sensory resolution. For example, even in the noisy condition above, if the range of sensory information decreases from [0, 1000] to [500, 550], then a sensory resolution of $\frac{550-500}{20} = 2.5$ units results.

Predictive coding achieves this reduction in sensory range by normalizing the original values by their predicted values, with predictions based on local averages in time and/or space. In the above five-integer example, the naive original range was [0, 1023] (hence the need for 10 bits per value), but normalizing by subtracting the average reduced the range to [−3, 3], which could be transmitted either with fewer bits or with the original 10 bits but now with much higher resolution.

Srinivasan, Laughlin and Dubs (1982) showed that, indeed, in the retina, predictions derived from local averages greatly reduced the salient sensory ranges, and thus the brain's relatively coarse-grained neural signals can still transmit high-resolution sensory information: small *differences that make a difference* (Edelman and Tononi 2000). For spatial filtering, the basic mechanism is lateral inhibition (see figure 5.2), whereby retinal neurons reduce one another's outputs via contributions to nearby interneurons, known as *horizontal cells*, which provide negative feedback to sensory receptors, that is, the rods and cones. Note that this computes something similar to a stimulus' spatial derivative (value minus its average surroundings), which is then sent downstream.

Achieving predictive coding across time requires a more local neural architecture: individual neurons (or pairs) can perform delayed feedback self-inhibition, essentially computing a temporal derivative to send further up the processing hierarchy. Consider the situation in figure 5.3. With each successive pass through the receptor loop, older values have less influence on the final output (since $|w| < 1$) but maintain a trace effect that diminishes more rapidly for smaller magnitudes of the inhibition gain, w. Note however that the alternating signs in $d - wc + w^2b - w^3c$ fail to model the difference between a current value (d) and a weighted sum of its past values. A more accurate model recruits the inhibitory interneuron as an accumulator that leaks a fraction $(1 - w)$ of its previous value while incrementing with

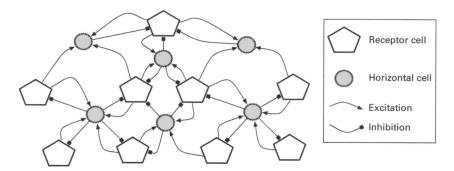

Figure 5.2
Basic architecture for spatial predictive coding in the retina, wherein horizontal cells aggregate inputs from nearby receptor cells and then inhibit the receptors with a strength commensurate with horizontal activity. The signal sent downstream by the receptors is then (approximately) their original input minus the neighborhood average (i.e., the prediction). In the actual retina, horizontal cells compute averages over thousands of receptors (Sterling and Laughlin 2015).

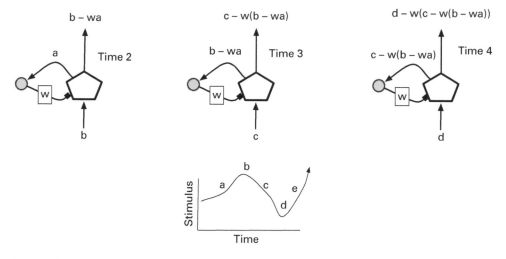

Figure 5.3
Basic synaptic connections required for temporal predictive coding, wherein the receptor cell (pentagon) inhibits itself via an interneuron (jagged gray oval). The output of the receptor at time t feeds back as a negative signal (with $|w| < 1$) at time t+1. The signal sent downstream by the receptor is then a fuzzy version of its temporal derivative.

the newest value. This accumulated amount then governs the strength of inhibition received by the primary receptor, as shown in figure 5.4. Now the output of the receptor manifests reality (the current value) minus a prediction (the weighted sum of previous values), with older values playing a less important role.

As discussed by several researchers (Stone 2018; Sterling and Laughlin 2015; Srinivasan, Laughlin, and Dubs 1982), predictive coding also has remedies for *extrinsic noise* in the original sensory stimuli (which reduces the signal-to-noise ratio of incoming signals). One key strategy is to increase sampling as the basis for predictions. For example, under conditions of low luminance, photoreceptors become particularly vulnerable to noise from random photons, since those photons may constitute a significant portion of the total photons received. This problem disappears in strong light, but to combat it in darker conditions, the predictive

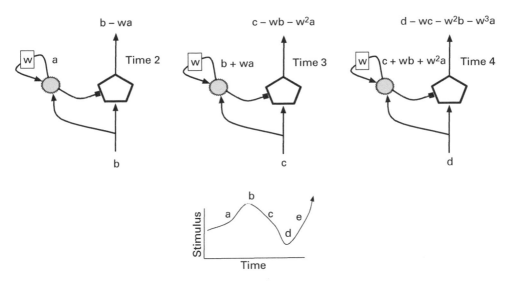

Figure 5.4
An improved model (compared to figure 5.3) of temporal predictive coding, with the interneuron accumulating the
stimulus history and then using it to normalize the current signal.

coder can increase sampling in space and/or time. In the spatial context, a cell can expand
the radius of neighbor cells that contribute to its prediction (i.e., local average), while temporally, it can increase the feedback gain (w) such that more historical values have significant
influence on the prediction. Although beyond the scope of this book, physicochemical mechanisms enable the retina to adjust parameters such as the feedback gain and effective radius
of spatial influence to properly adapt to widely varying lighting conditions and the diverse
signal-to-noise ratios that they incur (Sterling and Laughlin 2015).

This dynamic aspect of predictive coding has dimensions beyond those of the general ambient signal characteristics: the network can also adapt to predominant sensory
patterns, becoming less sensitive to the status quo and more responsive to *surprising*
inputs. As detailed by Hosoya and coworkers, plasticity of the synapses from inhibitory
interneurons to pattern-detecting cells can explain this flexibility (Hosoya, Baccus, and
Meister 2005). Figure 5.5 illustrates the general concept, with generic components (detectors and inhibitors) playing the roles of retinal ganglia and amacrine cells, respectively.
These reside a few levels downstream from the receptors and horizontal cells portrayed
in figure 5.2, but the general relationships between receptor / detector and inhibitor are
similar.

In this model, the activation of receptors and downstream detectors (having the former
within their receptive fields) will also activate local inhibitory cells, thus normalizing all
sensory signals by predictive feedback. Within regions affected by the stimulus (i.e., the
dark gray pattern in figure 5.5), both detectors and neighboring inhibitors will have high
firing rates, thus leading to Hebbian strengthening of the synapses between them. However,
this Hebbian tuning actually has an anti-Hebbian effect: future stimulation of the inhibitor
will cause greater *depression* of the detector. Thus, upon a later presentation of the same
pattern, the detectors will initially fire but then experience strong inhibition to significantly
weaken the standard predictive-coding signal (reality minus prediction) sent upward in the

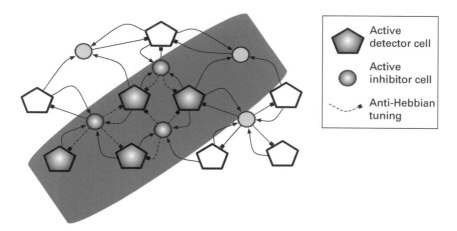

Figure 5.5
Generic model (based on relationships among bipolar, amacrine, and gangliar cells in the retina) of dynamic predictive coding that produces anti-Hebbian behavior via Hebbian modification of inhibitory synapses, as detailed in Hosoya, Baccus, and Meister (2005). The large dark pattern denotes the active receptive field to which downstream detector neurons (pentagons) respond, thus stimulating local inhibitor cells, which then negatively feed back on the detectors to manifest predictive coding. By modifying the synapses from the inhibitors to the detectors in a Hebbian manner, the predictive coding becomes dynamic: it exhibits lower sensitivity to the same (expected) pattern in the future. In this caricature, detectors are stimulated by the pattern, while inhibitors fire only in response to two or more active neighboring detectors.

neural hierarchy. The pattern will no longer be surprising and will thus stir up little activity beyond the earliest sensory levels.

In this manner, predictive coding becomes dynamic and can explain a wide range of adaptive effects in sensory perception. Note the difference between this and the *static* predictive coding discussed so far: the earlier variants involve predictions based on the immediate spatial or temporal situation. They can detect gradients / contrast, both weak and strong, by normalizing against the average. However, they learn nothing from the experience such that repetition of a sensory pattern minutes later will invoke the same levels of prediction error and thus be equally *surprising* to the network. In dynamic predictive coding, synaptic modification allows expectations to function across temporal scales much larger than the millisecond windows of neuronal firing; the novel pattern quickly becomes mundane and demands less signaling bandwidth when later repeated.

5.2   Predictive Coding on High

As the synaptic distance from sensory receptors increases in moving up the neural hierarchy—and thus as the receptive fields of individual neurons expand—the accuracy of predictions based on neighborhood averages should decrease and eventually disappear. Although topological maps (whereby neighboring neurons have neighboring receptive fields) do persist across many levels of sensory cortex (Kandel, Schwartz, and Jessell 2000), they vanish in higher cortical regions and the hippocampus, so the firing levels of neighboring neurons need not exhibit salient correlations. Similarly, as timescales for neural processing increase with distance from the sensory inputs and motor outputs, the utility of temporal derivatives for neural-activity prediction also diminishes.

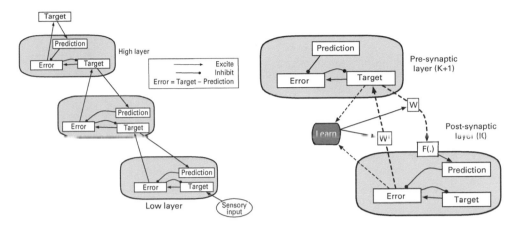

Figure 5.6
(Left) The essence of Rao and Ballard's (1999) predictive-coding model of the cortical hierarchy. (Right) Two neighboring levels of the same predictive-coding model in which weighted connections between targets and predictions are learned via Hebbian mechanisms based on correlations between the target at level K+1 and the error at level K. The (negated) predictions and targets at level K feed directly into the error units at K (in a one-to-one manner) without additional weighting. Each prediction, target, and error box represents a vector of twenty neurons for most of the simulations discussed in the text.

However, according to many contemporary models of cortical-column architecture and dynamics (Carpenter and Grossberg 2003; Mumford 1992; Rao and Ballard 1999; Spratling 2008, 2017), the brain has alternative predictive-coding mechanisms. Many of these shift the focus from shrinking the range of values encoded by individual neurons to reducing the sheer number of neurons in a cortical level whose signals require further (upward) transmission. Most of these models follow Rao and Ballard's (1999) basic paradigm, in which neurons in higher regions suppress the activity of lower-level neurons, whose outgoing signals now represent error in the sense of a mismatch between predictions and the *reality* conveyed by sensory signals. Rao and Ballard's work was inspired by David Mumford's (1992) theories about the functional significance of cortical columns and their connectivity patterns. Mumford's special focus was on the high degree of intercolumn linkage, in both the bottom-up and top-down directions, which portrays these columns not as modular units with simple, restricted interfaces and local computations, but as highly interactive groups whose bottom-up *residuals* (analogous to prediction errors) eventually reconcile with top-down predictions via an oscillating relaxation process similar to Carpenter and Grossberg's adaptive resonance theory (2003).

Figure 5.6 (left) portrays three predictive-coding levels (plus a topmost target vector, y_3) and the relationships among predictions, errors, and targets, where the latter denote sensory reality that predictions attempt to match. Note that the only signal sent upward is the error. Hence, whenever the target and prediction align, upward information flow attenuates; only error (aka *surprise*) need propagate further. Also note that the target at level K stems proportionately from the error at level K − 1 and inversely from the error at level K. Thus, errors at level K − 1 lead to modified targets at level K, which, in turn, alter the predictions at level K − 1.

Furthermore, when following the synaptic links around either of the main cycle motifs in the diagram (i.e., error-target-prediction-error and error-target-error), notice the presence

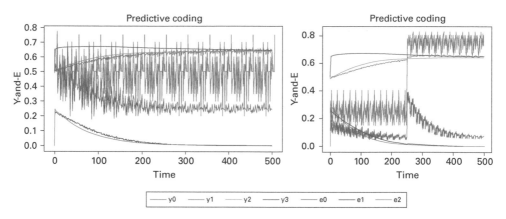

Figure 5.7
(Left) Progression of averages for target (y_i) and error (e_i) vectors, each of length 20, across three levels of a predictive-coding network similar to that of figure 5.6, where smaller indices denote lower levels and vector y_0 houses the time-varying inputs (that repeat every 50 timesteps). Input vectors are random uniform samples from set $\{0,0.5\}$. The plotted values of target vectors are arithmetic averages, while mean-square averaging applies to the error vectors; for each interlevel weight matrix, the learning rate λ is 0.001.(Right) Run of predictive coder where input fifty-vector sequence (y_0) changes at timestep 250 from samples of $\{0,0.5\}$ to samples of $\{0.5,1.0\}$.

of an odd number of inhibitory relationships, a prime indicator of negative feedback, which should encourage activation patterns to stabilize.

Figure 5.6 (right) displays a portion of the basic predictive coder in greater detail, revealing the weighted connections between the targets of level K+1 and the prediction (and then error) of level K. The jth neuron in the 20-unit prediction vector of level K receives weighted inputs from all 20 target neurons in level K+1. The weighted sum of these inputs then feeds into a sigmoid activation function to produce the jth prediction, which is then subtracted from the jth target value (also at level K) to yield the jth error term (at level K). Following the basic algorithm for a predictive coder (Spratling 2017), the transpose of the weight matrix that links targets (at level K+1) to predictions (at level K) is used to map the errors (at level K) to the target (at level K+1).

Adaptation in these networks occurs in two ways: (1) changes to targets in response to the errors at their own level and below, and (2) changes to *influences* of targets on errors (and vice versa) realized by the weight matrices. The latter constitutes learning and results from a standard Hebbian process: $\triangle w_{i,j} = \lambda e_i y_j$, with learning rate ($\lambda$), error ($e_i$) from level K, and target (y_j) from level K+1.

A simple simulation of the three-leveled network of figure 5.6 was performed using random input vectors of length 20, but whose sequence was repeated at intervals of 50 timesteps, thus giving the network an opportunity to adapt and improve. The results in figure 5.7 illustrate the transition to low errors and stability, despite the high variability of the input stream, y_0. In the leftmost plot, notice that all three target vectors (y0–y2) stabilize at nearly the same average, while the two higher-level error vectors (e1 and e2) converge to zero. Thus, by learning proper predictions, as embodied in both the prediction vectors and the weight matrices, the network drastically reduces the upward flow of error signals, with the only significant interchanges occurring between the lowest two levels. As shown in the rightmost plot, the predictive coder easily adapts to major changes in the composition of the input stream, with relatively large transitions at the lower levels (i.e., level-0 errors, e_0, and level-1 targets, y_1) but only barely perceptible changes further up in the hierarchy, thus indicating coarser, more general information encoded in the upper levels. Notably, the plots of figure 5.6 are almost identical to those in which a three-tiered neural PID controller (displayed in chapter 6) was run on the same data. This is not so surprising, given the abundance of negative feedback loops mentioned above.

The general dynamics of primitive predictive-coding models such as that of figure 5.6 mirrors the behavior of cortical networks proposed by Jeff Hawkins in his groundbreaking book, *On Intelligence* (Hawkins 2004), and related publications (Hawkins and Ahmad 2016; Hawkins, Ahmad, and Cui 2017). Hawkins uses detailed analyses of the six-layered cortex to explain how brains could propagate sensory information upward only until it meets the downward flow of matching expectations: predicted sensory realities need not disturb higher cortical regions. Only surprising information travels far, possibly to the hippocampus, which represents the pinnacle of the cortical hierarchy and the gateway to long-term memory formation.

Along with Hawkins, several renowned systems neuroscientists (Rodriguez, Whitson, and Granger 2004; Ballard 2015) employ similar network models to explain the neocortex, particularly with respect to perception and prediction. These models draw on thorough accounts of cortical neuroanatomy from Vernon Mountcastle's seminal book, *The Cerebral Cortex* (Mountcastle 1998). The essence of these models is summarized in figure 5.8, which depicts three six-layered cortical columns. The following descriptions of the cortical column draw from those of Mountcastle, as well as Rolls and Treves (1998), Ballard (2015), Schneider (2014), and Hawkins (2004).

The basic design of these columns is preserved across the entire cortex and across species as diverse as mice and humans. Layer 1 consists almost exclusively of dendrites serving the excitatory pyramidal cells in all other layers, but particularly 2, 3, and 5. Bottom-up signals normally exit the lower-level column from layer 2/3 and enter the higher column at level 4, which contains primarily excitatory stellate cells. These resemble pyramidals but tend to have shorter axons and only synapse locally, in layer 4 and with the layer-2/3 pyramidals, which then convey signals both upward to higher columns and down to layers 5 and 6 of the same column. Top-down signals move primarily from layers 5 and 6 of the upper column to both the thalamus and the layer-1 dendritic mats of the lower column, then through layer 2/3 and on to layers 5 and 6 for further transmission down the hierarchy.

Although outnumbered by excitatory cells by a ratio of approximately five to one in the cortex, inhibitory neurons also play a vital role in columnar behavior: they help sparsify the activity patterns of the column's excitatory cells. The primary inhibitory neurons are basket and chandelier cells, which have short spatial ranges of influence of no more than 100 micrometers, whereas projections from excitatory neurons may extend several millimeters (Kandel, Schwartz, and Jessell 2000). Inhibitors often synapse directly on the somas or axon hillocks of excitatory neurons, thereby exerting a strong blocking effect. The presynaptic terminals of an inhibitor's axons emit GABA, whose receptors on postsynaptic terminals have slower binding and release times than those found at receptors for the excitatory neurotransmitters such as glutamate and AMPA. Thus, whereas the excitation of a pyramidal cell lasts 15–20 milliseconds, its inhibition normally endures for 100–150 msec (Rodriguez, Whitson, and Granger 2004).

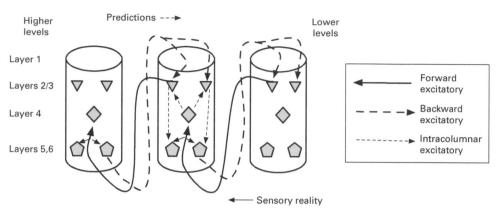

Figure 5.8
Basic functional circuits in the neocortex showing only the main intra- and intercolumnar excitatory connections. Columns are differentiated functionally (by their receptive fields), not necessarily anatomically, with each containing tens of thousands of pyramidal (pentagons and triangles) and stellate (diamonds) cells, grouped into modules housing hundreds of vertically interacting neurons in layers 2–6. Layer 1 is essentially a dendritic mat (Mountcastle 1998; Rodriguez, Whitson, and Granger 2004; Schneider 2014).

Lateral inhibition is commonplace in the brain, such that when a neuron N fires, it excites local inhibitors that quell the activity of neighboring neurons and (after 15–20 msec) N itself, all for the extended interval of 100–150 msec. This serves at least two purposes: (1) immediate hampering of neighbors promotes sparse activation patterns, which are much more convenient for information transmission and storage; and (2) strong, enduring inhibition of the originally active neuron(s), which constitute an activation pattern, gives other neurons an opportunity to participate in the next pattern, which helps reduce the overlap (and possible interference) between activation states, an obvious advantage for information storage and retrieval in a distributed memory.

The precise anatomical and physiological details of cortical columns (Mountcastle 1998; Schneider 2014; Rodriguez, Whitson, and Granger 2004) are far beyond the scope of this book, but some have special relevance for predictive coding, despite the lack of a fully comprehensive and empirically validated theory. In particular, the role of inhibition seems critical, since predictive coding entails a weakening of bottom-up signaling when top-down predictions match the rising sensory patterns. How could higher levels inhibit lower levels?

First of all, the short spatial range of inhibitory neurons indicates that level K+1 probably inhibits level K by sending down excitatory signals that trigger local inhibitors in level K. At this point, the difference between bottom-up and top-down signals becomes important. The former typically enter a column directly via layer 4, with *proximal* synapses (i.e., those near the cell soma), and with many layer-4 pyramidals receiving similar axonal input. Thus, they all have similar receptive fields. Conversely, the predictive action potentials enter via the dendritic mats of layer 1, thus synapsing *distally* (i.e., far from the soma) and therefore having a much weaker effect on the soma; a single top-down signal cannot normally induce the soma to generate action potentials, though it can lead to some degree of depolarization.

Hawkins and Ahmad (2016) leverage this difference to provide an interesting explanation of the essence of predictive coding: the weakening of upward signals when predictions

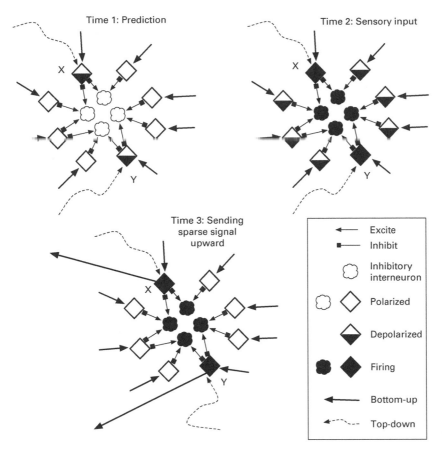

Figure 5.9
Key mechanism underlying predictive coding as hypothesized by Hawkins and Ahmad (2016). Time 1: Predictive signals from a higher level weakly excite (depolarize) a few level-4 excitatory neurons (diamonds). Time 2: Sensory inputs from a lower level begin exciting neurons X, Y, and their neighbors. X and Y begin firing before their neighbors, thereby exciting the inhibitory neurons, which repolarized the neighbors, leaving only X and Y active enough to send their signals upward at Time 3.

match sensory signals. As shown in figure 5.9, the predictive signals coming from a higher level via layer-1 dendrites provide only weak stimulation to a few excitatory level-4 neurons. This depolarizes those cells to some degree, but not enough to invoke action potentials. However, as the bottom-up sensory signals begin coming in, these cells become fully depolarized ahead of their neighbors. The ensuing action potentials then activate neighboring inhibitory neurons (e.g., basket cells), which begin suppressing nearby neurons such that many of them never reach firing thresholds, as classic winner-take-all lateral inhibition. This sparse population of *winners* then sends signals upward via layer 2/3. In the absence of a biasing predictive signal, the set of winner cells could be much larger, thus requiring higher bandwidth to continue propagating the bottom-up signal.

One key detail abstracted away from figure 5.9 is the relationship between layer-4 excitatory cells. These are often spiny stellate cells (not pyramidals) which have short axons but which tend to excite many of their neighbors. Hence, each diamond in the figure represents a small population of mutually stimulating spiny stellate cells. Once activated to firing thresholds, these cells would autocatalytically sustain one another despite inhibitory interference, whereas the slower-charging stellates would never reach that level, remain vulnerable to inhibition, and never become an active cell assembly.

Another, equally plausible, explanation focuses on inhibition at level 2/3. If the predictive signal activates pyramidals in layer 2/3 (again via the layer-1 dendritic mat), the prospects for depolarizing those neurons to threshold are better than for level-4 neurons, since the synapses into the dendritic mat should be closer to the layer-2/3 somas (than to layer-4 somas). Assuming that some of these level-2/3 neurons begin firing, their assemblies may remain active for 15–20 msec before succumbing to inhibition for 100–150 msec. During that inhibitory period, many level-4 stellate assemblies could be driven to firing by bottom-up signals. These level-4 signals must then go through layer 2/3 on their journey upward, but all routes through the inhibited layer-2/3 neurons will be blocked. Thus, only those level-4 assemblies *not* matched by a predictive signal could continue up the hierarchy, assuming that their corresponding neighborhoods in level 2/3 reside far enough away from the active pockets of inhibition. This further illustrates the advantages of pattern sparsification: a sparse sensory signal can more easily slip past (completely intact) the inhibitory zones induced by an erroneous prediction.

In general, one important fact should be kept in mind when discussing theories of cortical columns and their functionality: size. The neuroscience literature (Mountcastle 1998; Hawkins 2004; Hawkins, Ahmad, and Cui 2017; Amit 2003; Rakic 2008) varies on the estimates, but the primate brain probably contains on the order of 10^5–10^6 columns, with each consisting of a few hundred mini columns, each, in turn, housing around one hundred, tightly interconnected neurons representing a majority of the main neural cell types. Hence, a cortical column consists of several tens of thousands of diverse neural cells. The neurons in each minicolumn tend to have highly correlated firing patterns, while a full column appears to handle a particular function associated with a receptive field, motor region, or cognitive task. The combination of thousands of diverse units provides plenty of computational machinery for realizing a broad spectrum of these functions. Theories abound as to their identity, but few should be ruled out simply on the grounds of complexity, since each column seems to *have the numbers* and diversity to perform any of a wide range of calculations, with variables involving considerable spatial and temporal scope. The basic mathematical operations discussed in earlier chapters, including those of a PID controller, could easily be handled by this cellular machinery.

5.2.1   Learning Proper Predictions

Descriptions of predictive coding often assume preexisting neural circuitry that links representations of *causes* at one level with those of predicted *effects* in the level below, possibly with a brief nod to Hebbian mechanisms as the underlying generative principle. Unfortunately, there is no consensus explanation of how these circuits emerge through a combination of development and synaptic tuning. What follows is one possible explanation based on several published theories (Downing 2009, 2015; Hawkins 2004; Rodriguez, Whitson, and Granger 2004; Wallenstein, Eichenbaum, and Hasselmo 1998).

As expected, the story begins with Donald Hebb and neurons that *fire together, wire together* (Hebb 1949; Fregnac 2003), but details of spike-timing-dependent plasticity (STDP) also come into play. As summarized by Song and colleagues, the firing-time

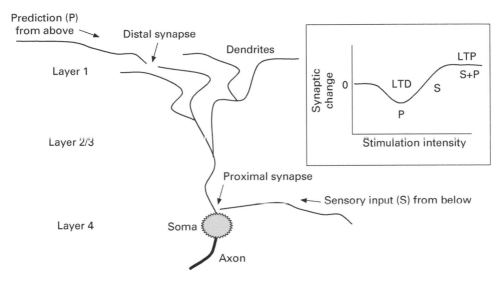

Figure 5.10
Basic neurocytology in the cortical column, layer 4 (aka the granular layer), with afferents from lower levels (e.g., sensory areas) synapsing proximally (i.e., close to the soma) and top-down predictive signals synapsing distally (i.e., far from the soma) in the dendritic mat of layer 1. (Inset) Differences in synaptic change for low- versus high-stimulation scenarios, where weak input (in the form of an unmatched prediction) leads to depression (LTD), while coincidental predictive and sensory input creates sufficient activation to produce synaptic potentiation (LTP).

relationships between presynaptic neuron A and postsynaptic neuron B play a critical role in synaptic modification (Song, Miller, and Abbott 2000). If A's spikes occur just before B's, then the $A \rightarrow B$ synapse potentiates (i.e., strengthens), but if B spikes before A, that synapse depresses (i.e., weakens). Related work by Artola, Brocher, and Singer (1990) shows that a postsynaptic neuron experiencing high stimulation will potentiate its afferent synapses (those associated with its dendrites), while weak stimulation tends to depress those same synapses. This creates interesting dynamics in cortical columns.

Looking at the layer-4 neuron in figure 5.10, note that afferents from lower levels (e.g., sensory inputs) tend to synapse very close to the soma, whereas afferents from higher levels (carrying predictive signals) synapse distally in the dendritic mat of layer 1. Action potentials arriving at these distal synapses have relatively long distances to travel, and hence attenuate before reaching the soma. Thus, whereas bottom-up signals can *drive* a layer-4 cell, top-down signals, on their own, can provide only weak stimulation, at least until the distal synapse has been enhanced by LTP.

As the inset of figure 5.10 indicates, the weak signal provided by an isolated predictive signal (P) may produce LTD, as implied by Artola, Brocher, and Singer (1990), but the combination of sensory input (S) and P should stimulate the soma enough to induce LTP at both the proximal and distal synapses. If potentiated enough, the distal synapses should eventually be capable of delivering strong predictive signals to the soma, thus producing considerable spiking activity even in the absence of the bottom-up signal. Conversely, a predictive signal that goes unmatched by a sensory signal may produce LTD and thus reduce the future influence of top-down signals. These mechanisms could play a crucial role in the formation of predictive circuits.

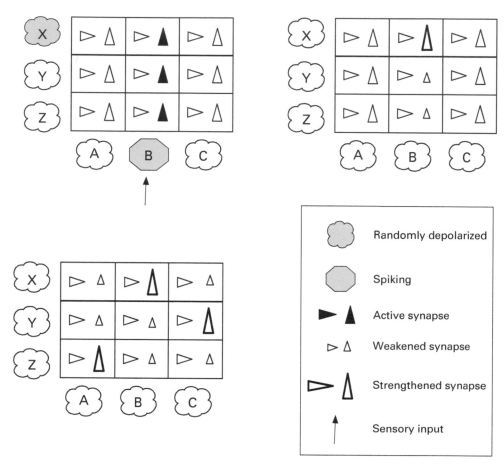

Figure 5.11
Basic mechanism by which bottom-up connections can form during early learning and development. Neurons A,B, and C are one level (L1) below neurons X,Y, and Z (level L2). Bottom-up connections denoted by vertical triangles/arrows; top-down connections by horizontal triangles/arrows. Initially, all interlevel connections exist and have similar efficacies. (Upper left) Sensory input stimulates B, which activates all $B \rightarrow$ L2 synapses (dark vertical arrows), while X randomly depolarizes (due to typical random activity during development). (Upper right) Coincident activity of B and X leads to long-term potentiation (LTP) of the $B \rightarrow X$ synapse, while $B \rightarrow Y$ and $B \rightarrow Z$ weaken via long-term depression (LTD) due to active presynaptic but inactive postsynaptic neurons. (Lower left) Similar combinations of sensory input to A (and C) and tonic activity of Z (and Y) produce two more biased connections: $A \rightarrow Z$ and $C \rightarrow Y$.

Although the descriptions that follow involve individual neurons, this is primarily for illustrative purposes. The true story surely involves neural assemblies, and each of the units discussed below should probably be interpreted as a collection of cooperating cells.

Two different pathways need to be established: the driving circuits from lower to higher levels, and the predictive networks from higher to lower levels. Figure 5.11 gives a procedure for tuning the driving synapses from level L1 to L2. Two important factors are (1) an initial network in which most L1 neurons connect (via proximal synapses) to most L2 neurons, and (2) the random depolarization of L2 neurons, a common event in the brain, particularly during development. Thus, an L2 neuron (X in the figure) that randomly activates slightly after an L1 neuron (B) will promote LTP on the proximal $B \to X$ synapse. In addition, the L2 neurons (Y and Z) that fail to activate in that same time frame will experience depression on their synapses from B.

In Hebbian learning, presynaptic firing that is not followed by postsynaptic activity also leads to synaptic depression. During this phase, the activity in L2 may involve depolarization without actual spiking, such that the synapses from L2 to L1 are not affected. However, once the bottom-up proximal connections become fortified by this first phase, future activation of neurons such as D will suffice to depolarize X up to its spiking threshold.

In phase 2 (see figure 5.12), the L2 neurons can now produce significant action potentials and exert some influence on L1. However, the distal nature of all connections from L2 to L1 presents challenges for any predictive signal, which must first traverse the unrefined synapses of the layer-1 dendritic mat. This is where Artola and colleague's findings come into play. When an L2 neuron (e.g., Z) sends an action potential across a layer-1 synapse and down to a layer-4 neuron (e.g., B) in the same 20–40 msec time window that B receives a driving signal from a lower level, B will fire intensely and the $Z \to B$ synapse (in the dendritic mat) should experience LTP, thus forming a strong link between Z and B. When Z fires, it will also send signals across the $Z \to A$ and $Z \to C$ synapses, but since neither A nor C receives a bottom-up signal in the same short window, A and C will be only weakly stimulated, leading to depression of the $Z \to A$ and $Z \to C$ synapses. In this way, Z becomes a predictor of B. Anthropomorphically speaking, Z *searches* for matching bottom-up signals by trying all post-synaptic possibilities and then potentiating where it finds matches and depressing where it does not.

Refinement of the top-down synapses proceeds as explained in figure 5.12, with the L2 neurons serving as links between neurons A, B, and C in L1. Although seemingly superfluous in this simple example, these higher-level intermediaries become vital when A, B, and C actually represent cell assemblies in a shared cortical region. Creating strong intralevel links between many members of distinct assemblies could easily create overconnected levels in which almost all neurons could become simultaneously active, thus destroying the information-encoding capacity of that level. Adding the extra level provides useful time delays and dedicated high-level detectors for and stimulators of individual assemblies at the lower level.

Figures 5.11 and 5.12 show the formation of a network for remembering the sequence A-B-C, but how does this scale up to hierarchies in which patterns at upper levels represent abstractions of lower-level patterns? The key lies in timescale differences across the neural hierarchy: higher cortical regions run at slower timescales that lower regions (Raut, Snyder, and Raichle 2020). Thus, when a lower-level driving signal activates a higher-level neuron/assembly, the latter remains active (high spiking) longer than the former. This would allow the higher-level pattern to bind to several lower-level patterns.

Returning to figure 5.12, after neuron/assembly A activates Z, the latter may remain active throughout the bottom-up stimulation of B, C, and several other units/assemblies. Thus, in the future, Z would trigger on A but then predict B, C, and so on, thereby representing chunked sequences. The other intermediate neurons (X and Y), though originally molded as detectors of B and C, become superfluous (and thus recyclable as detectors for other patterns). When Z predicts B and C, the matching of expectations to reality (the bottom-up signals for B and C) reduces the upward signaling (as described earlier and shown in

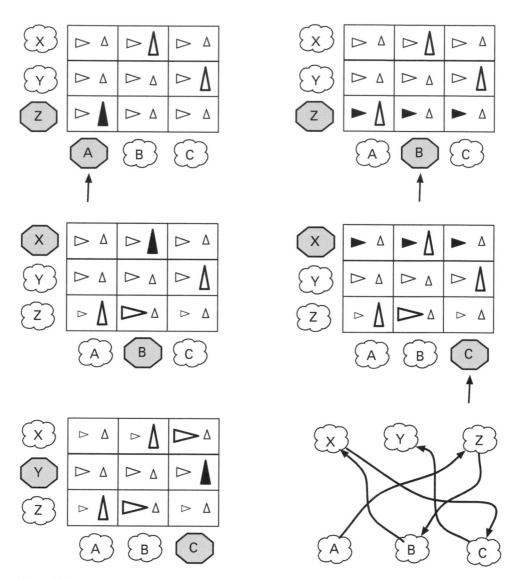

Figure 5.12
Basic mechanism for refining top-down connections during learning, with the same legend as in figure 5.11. (Top Left) L1 neuron A receives bottom-up stimulation on a proximal synapse, thus sending an action potential across the $A \to Z$ synapse and exciting neuron Z. (Top Right) Z's firing sends signals across all three synapses to L1 in the same short time window that neuron B receives a driving, bottom-up signal. (Middle Left) This strengthens the $Z \to B$ synapse but weakens those to A and C (neither of which receive driving signals in this time window). Activated B then stimulates X via the previously enhanced $B \to X$ synapse. (Middle Right) Spiking X sends action potentials down all synapses to L1 at the same time as a driving bottom-up signal reaches neuron C. (Bottom Left) This potentiates the $X \to C$ synapse but depresses $X \to A$ and $X \to B$. (Bottom Right) Summary of the strong synapses that realize the activation chain: $A \mapsto Z \mapsto B \mapsto X \mapsto C \mapsto Y$.

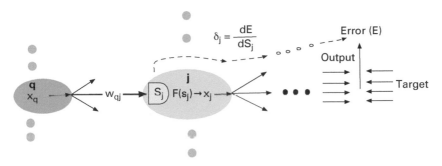

Figure 5.13
Basic relationships among neurons (gray ovals), the sum of their weighted inputs (S), their activation function (F) and output (x), and their δ gradient, $\frac{\partial E}{\partial S_j}$, which links the sum of weighted inputs S_j of any node (j) to the error (E) in the output layer, which may be many layers downstream from node j.

figure 5.9) such that X and Y no longer become activated. In this way, neurons at higher levels gradually adapt to become abstractions (aggregates) of lower-level details.

5.3   Predictive Coding for Machine Learning

Artificial learning systems for both simple and relatively complex tasks do not require the complexity of STDP and cortical columns, as contemporary machine learning (ML) successes have clearly shown. However, the essence of predictive coding provides a more biologically plausible (and still highly effective) alternative to ML's classic supervised learning tool: the backpropagation network. To understand the potential contributions of predictive coding to ML, a brief review of the backpropagation algorithm is in order.

5.3.1   The Backpropagation Algorithm

At the core of the algorithm are gradients, many of them of the long-distance variety, expressing the influence of various system variables on the objective function (aka cost or error function). For standard backprop calculations, the two main types of gradients are $\frac{\partial E}{\partial w}$ and $\frac{\partial E}{\partial S}$. The former expresses the effect of a synaptic weight on the output error, while the latter denotes the effect of the sum-of-weighted-inputs (S) to a node on that same error. The former is often calculated from the latter, which mainly serves as an intermediate variable, δ, but which supplies the key link to predictive coding (described below).

Figure 5.13 provides many of the important ties between variables, with node j as the focal point. Note that its sum of weighted inputs, S_j passes through an activation function (F) to produce the node's output value (x_j). Further note that upstream neighbor node q sends its output to node j along the connection with weight (strength) w_{qj}. Thus, two simple calculations reveal that

$$\frac{\partial S_j}{\partial w_{qj}} = x_q \tag{5.1}$$

and

$$\frac{\partial S_j}{\partial x_q} = w_{qj} \tag{5.2}$$

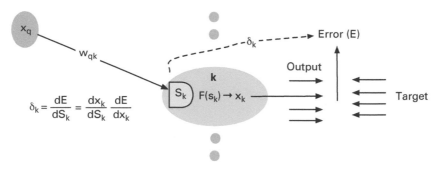

Figure 5.14
Computation of the influence of an output node's sum-of-weighted-inputs (S_k) on the error (E): $\delta_k = \frac{\partial E}{\partial S_k}$.

The derivative of this sum with respect to both the weight and the upstream output are important parts of the gradient chain.

Given δ_j for any node j, the chain rule of calculus easily produces $\frac{\partial E}{\partial w_{ij}}$ for any upstream neighbor node i:

$$\frac{\partial E}{\partial w_{ij}} = \frac{\partial S_j}{\partial w_{ij}} \frac{\partial E}{\partial S_j} = x_i \delta_j \tag{5.3}$$

And this gradient provides the essential information for intelligent weight change using the standard update rule (with λ as the learning rate):

$$\Delta w_{ij} = -\lambda \frac{\partial E}{\partial w_{ij}} \tag{5.4}$$

In short, given a node's δ gradient, the gradients for all incoming weights are trivially computed. Calculating the δ gradients themselves is a bit more complicated. The simplest case is on the output end of the network, as illustrated in figure 5.14, where the link from S_k to error involves one intermediate term, x_k, the output of neuron k, also a network output.

Adding a little more detail, assume a standard mean-squared error function:

$$E = \frac{1}{2} \sum_i (T_i - x_i)^2 \tag{5.5}$$

where T_i is the target value for the ith output neuron. Then, for any particular output neuron, k:

$$\frac{\partial E}{\partial x_k} = -(T_k - x_k) \tag{5.6}$$

and hence:

$$\delta_k = \frac{\partial E}{\partial S_k} = \frac{\partial x_k}{\partial S_k} \frac{\partial E}{\partial x_k} = -F'(S_k)(T_k - x_k) \tag{5.7}$$

Thus, the gradient for weight w_{qk} on the connection from upstream neighbor node q to node k (in figure 5.14) is

$$\frac{\partial E}{\partial w_{qk}} = x_q \delta_k = -x_q F'(S_k)(T_k - x_k) \tag{5.8}$$

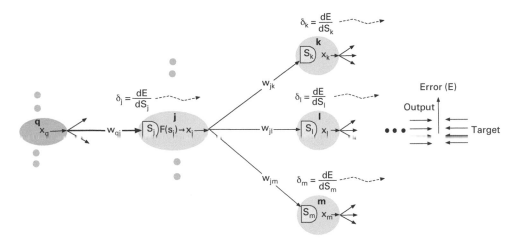

Figure 5.15
Illustration of the recursive relationship between the δ gradients across multiple layers of a neural network. As expressed in equation 5.11, the gradient at node j is the derivative of the activation function, $F'(s_j)$, multiplied by the sum of the weight-modified δ gradients of all immediate downstream neighbors.

And the weight update becomes[1]

$$\triangle w_{qk} = -\lambda \frac{\partial E}{\partial w_{qk}} = \lambda x_q F'(S_k)(T_k - x_k) \tag{5.9}$$

In networks with one or more hidden layers, the calculation of the non-output δ gradients requires recursion, wherein the gradient for one node depends on the gradients of all of its downstream neighbors. Figure 5.15 portrays that section of a multilayered network most relevant for computing the δ gradient for node j, in the middle of the diagram.

The chain rule breaks this down into two derivatives, one simple and one more complex:

$$\delta_j = \frac{\partial E}{\partial S_j} = \frac{\partial x_j}{\partial S_j} \frac{\partial E}{\partial x_j} = F'(S_j) \frac{\partial E}{\partial x_j} \tag{5.10}$$

Finding $\frac{\partial E}{\partial x_j}$ requires a summation over all paths taken by x_j in the network:

$$\delta_j = F'(S_j) \frac{\partial E}{\partial x_j} = F'(S_j) \left[\frac{\partial S_k}{\partial x_j} \frac{\partial E}{\partial S_k} + \frac{\partial S_l}{\partial x_j} \frac{\partial E}{\partial S_l} + \frac{\partial S_m}{\partial x_j} \frac{\partial E}{\partial S_m} \right]$$

$$= F'(S_j) \left[w_{jk}\delta_k + w_{jl}\delta_l + w_{jm}\delta_m \right] \tag{5.11}$$

And thus the update for incoming weight w_{qj} is

$$\triangle w_{qj} = -\lambda \frac{\partial E}{\partial w_{qj}} = -\lambda x_q \delta_j = -\lambda x_q F'(S_j) \left[w_{jk}\delta_k + w_{jl}\delta_l + w_{jm}\delta_m \right] \tag{5.12}$$

Note that the intermediate expression above,

$$\triangle w_{qj} = -\lambda x_q \delta_j \tag{5.13}$$

hints of a Hebbian or anti-Hebbian update rule, assuming that δ_j constitutes a local property of node j. This locality is definitely not the default case with standard backpropagation, since the δ gradient often represents the accumulation of many layers of downstream gradients, all with the same timestamp corresponding to the current input-output case. However, as shown below, the prediction error inherent in predictive coding is a local variable that behaves very similar to the δ gradient.

The name *backpropagation* stems from this recursive transfer of δ gradients backward, from output to input layer. The complete algorithm consists of a forward phase, where inputs are propagated through the net to the output layer, and a backward phase, where gradients and then weights are updated. The forward phase has no conflicts with neuroscience, and none of the individual calculations of the backward phase (e.g., gradient derivations) seems particularly difficult for a brain. But the caching of numerous activation values and the lockstep reverse traversal of the network (enabling recursion) has no neuroscientific basis. The brain is full of backward connections, but they are hardly symmetric to the forward links, either in presence or in synaptic strength (weight). Hence, any weights that modulate top-down (backward-passed) information in the brain will not precisely mirror the weights used during bottom-up (forward) processing along that same pathway, as backpropagation's calculations require. In general, the complete process of the backward phase requires a large stretch of the imagination to put in a biological context.

5.3.2   Backpropagation via Predictive Coding

Predictive coding circumvents these problems by essentially employing signals analogous to δ gradients during all phases of operation, producing continuous streams of that information between neighboring neurons, which can then use it to update both their activations and weights. These δ's embody prediction errors, denoted here as ξ's, which gradually accumulate influence from more distant network regions during the transition to equilibrium that characterizes normal activity in a predictive-coding network. Although not mathematically equivalent to δ gradients, the ξ's tend to converge to similar values (Whittington and Bogacz 2017) when performing supervised learning.

Figure 5.16 displays a general predictive-coding network similar to that of Rao and Ballard (1999), although their model assumes that each weight matrix (U) on a bottom-up connection is the transpose of the matrix W on the corresponding top-down link. To understand the relationship between backpropagation and predictive coding, it helps to simplify this diagram by removing the prediction boxes and feeding the weight-adjusted targets directly into the error nodes of their lower neighbor. The simplified model (see figure 5.17) accentuates the short negative-feedback (i.e., stabilizing) loops (a) between the error term and target of each layer, and (b) between the target of one layer and the error of the layer below. The following explanation of predictive coding based on figure 5.17 closely mirrors those given by Wittington and Bogacz in two detailed research papers on the topic (2017, 2019).

Target nodes act as integrators in predictive coding, whereas error terms represent the current difference between those targets and top-down predictions. Hence, the following equations express the dynamics of each:

$$\frac{\partial x^k}{\partial t} = U^k \xi^{k-1} - \xi^k \tag{5.14}$$

$$\xi^k = x^k - \underbrace{W^k x^{k+1}}_{prediction} \tag{5.15}$$

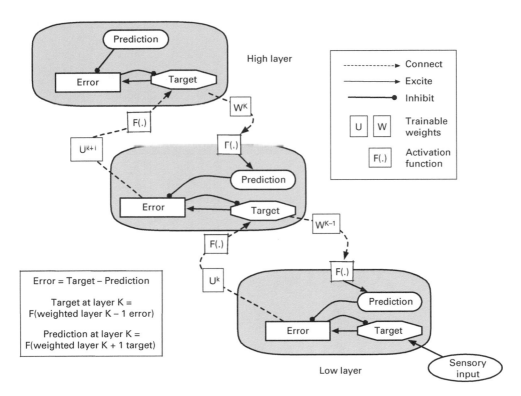

Figure 5.16
Detailed model of a predictive-coding network.

Now, consider the network of figure 5.17 running in both an unclamped (predictive or dreaming) mode and a clamped (recognition) mode. During predictive runs, no sensory input exists, thus leaving x^{k-1} (e.g., x^0) unconstrained from below. This lack of bottom-up restrictions combines with the negative feedback loop between ξ^{k-1} (e.g., ξ^0) and x^{k-1} to enable ξ^{k-1} to alter x^{k-1} so as to achieve a balance between x^{k-1} and the prediction entering ξ^{k-1} from above. In short, it allows the lowest-level prediction error to approach zero. This, in turn, essentially eliminates the bottom-up constraint of ξ^{k-1} on x^k, the target at the next layer, and this allows ξ^k to approach zero. Of course, doing so causes changes to x^k, which modifies ξ^{k-1}, but this lowest-level error node always has free rein over x^{k-1} and can alter it to push ξ^{k-1} back toward zero, thus *absorbing* any predictive perturbations from above.

This same process gradually propagates up the layer hierarchy, with corresponding adjustments filtering downward. Eventually, the prediction errors at all levels approach zero, and thus the left side of equation 5.15 vanishes for all k. This yields

$$x^k = W^k x^{k+1} \tag{5.16}$$

which is the standard relationship between the activations at one level and those at a neighboring level in a backpropagation network. For predictive coding networks using a nonlinear activation function (F), the same basic relationship holds, with equations 5.15 and 5.16 rewritten (respectively) as

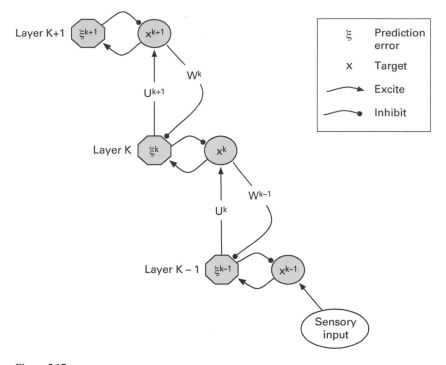

Figure 5.17
A simplification of the predictive-coding model of figure 5.16. Targets (x) at one level produce predictions at the level below when multiplied by weights W; the prediction then inhibits the error node at the lower level. Conversely, error terms (ξ) feed upward, across weights U, to excite the target node above.

$$\xi^k = x^k - \overbrace{W^k F(x^{k+1})}^{prediction} \tag{5.17}$$

$$x^k = W^k F(x^{k+1}) \tag{5.18}$$

Since all of the errors approach zero during predictive mode, their relationships are uninteresting. However, during clamped / recognition mode, all of the target vectors have dueling constraints and are not guaranteed to find equilibrium values that remove all error. Instead, the only reliable assumption is that the network will attain some degree of equilibrium, meaning that the left side of equation 5.14 vanishes, leaving the following relationship among the error vectors at neighboring layers:

$$\xi^k = U^k \xi^{k-1} \tag{5.19}$$

Taken together, equations 5.16 and 5.19 characterize a network that propagates activations in one direction while propagating errors in the opposite direction. A comparison of equations 5.19 and 5.11 reveals a key similarity between δ gradients and prediction errors (ξ): both manifest recursive properties by accumulating weighted versions of their neighboring counterparts.

Hence, predictive coding mirrors backpropagation in both (a) the dual flow directions for driving signals and error feedback, and (b) the recursive relationship between gradient (error) signals. Furthermore, note in equation 5.7 that the δ gradient at the output level is very similar to the output error. Thus, both backpropagation and predictive coding begin with similar types of values, which they then recursively combine backward through the network. The main difference between the algorithms is procedural: backpropagation performs paired forward and backward phases, with weight updates after each pair, whereas predictive coding runs with or without external targets through any combination of forward and backward propagation waves until reaching equilibrium, at which point weight updates occur.

Another convincing similarity stems from the weight updates. Predictive coding employs these simple Hebbian rules[2] (with learning rate λ):

$$\triangle U^k = -\lambda \xi^{k-1} x^k \tag{5.20}$$

$$\triangle W^k = -\lambda x^{k+1} \xi^k \tag{5.21}$$

Compare these to the update rule for backpropagation in equation 5.13, while keeping in mind the strong similarities between δ and ξ, and the two learning algorithms seem nearly identical. The big difference is that the δ gradient consists of a string of products and sums of *recent* activations, weights, and activation-function derivatives spread across large stretches of the network space, whereas the prediction error (ξ) represents an equilibrium value attained over time and also influenced by values across the network.

Predictive coding maintains locality by having explicit error neurons in each layer, but since these neurons also encode sensory and/or predictive signals, they are a natural part of the network. Conversely, δ gradients in backpropagation serve exclusively as complementary computational scaffolding for learning. When it comes to biological plausibility, that entire scaffolding network comes into serious question, as does the process needed to run it. Predictive coding, on the other hand, abides by a very emergent, biological mechanism: widespread bottom-up and top-down interactions leading to briefly stable activation patterns, with the main *missing biological link* being the existence of error nodes. Although nonconclusive, several neuroscientific studies (summarized by Kok and Lange 2015) do indicate the presence of distinct neuron groups that encode prediction error.

Two of the main researchers to reconcile predictive coding and backpropagation, James Whittington and Rafal Bogacz (2017, 2019), have achieved very high performance on supervised-learning benchmark data sets when using predictive-coding networks. A thorough analysis of their nets in action does indeed reveal a tight similarity between the values of (a) internal target activations (x^k) versus normal feedforward activations of a backprop net, and (b) weight updates (driven by δ and ξ, respectively) in backpropagation versus predictive-coding nets. For an overview of other promising, biologically plausible, backpropagation algorithms, see Lillicrap et al. (2020).

5.4   In Theory

The logic behind predictive coding seems rather impeccable. Why would the organ that punches way above its weight in terms of energy usage[3] waste any of that wattage to shuttle around superfluous information? When a neighbor passes my office window en route to our front door, waves, and sees me getting up and moving toward the entryway, should he knock to signal his arrival? Although common courtesy says *yes*, thermodynamics and information theory say *no, never; are you kidding me?*

Considerable neuroscientific and psychological evidence (summarized in Ouden, Kok, and Lange 2012; Spratling 2019; Clark 2016) supports both the predictive nature of many top-down signals and the use of prediction error as a common currency for bottom-up information exchange between many brain areas. These reports span the breadth of cranial faculties, from visual, auditory, and tactile perception to motricity to memory and language to motivational and cognitive control.

These myriad studies indicate that predictive coding could be prevalent throughout the nervous system. Whether or not a few common mechanisms (similar to those detailed in this chapter) can account for these diverse instantiations of the phenomenon remains debatable. The links between structure and function in the brain seem to have no desire to gratuitously share their secrets with scientists.

However, from the perspective of artificial intelligence, the sheer bulk of reputable natural-science studies in this area should motivate a deeper inquiry into the possibilities for synthetic, prediction-oriented neural networks. Fortunately, several prominent AI researchers got the message decades ago: they realized that prediction (in the form of pattern generation) plays a central role in perception, as a complement to bottom-up pattern recognition. Their neural networks were the topic of the previous chapter, while the work of Whittington and Bogacz (elaborated above) shows that predictive coding, with learning driven strictly by local gradients, can replace conventional backpropagation, at least for standard classification tasks.

Before moving on, a parting word by one of predictive coding's pioneers is in order (Attneave 1954, 192):

> The foregoing reduction principles make no pretense to exhaustiveness. It should be emphasized that there are as many kinds of redundancy in the visual field as there are kinds of regularity or lawfulness; an attempt to consider them all would be somewhat presumptuous on one hand, and almost certainly irrelevant to perceptual processes on the other. It may further be admitted that the principles which have been given are themselves highly redundant in the sense that they could be stated much more economically on a higher level of abstraction. This logical redundancy is not inadvertent, however: if one were faced with the engineering problem suggested earlier, he would undoubtedly find it necessary to break the problem down in some manner such as the foregoing, and to design a multiplicity of mechanisms to perform operations of the sort indicated.

In this final paragraph of his classic paper, Attneave speaks presciently to the nagging dilemma of every bio-inspired AI practitioner: which biological details to include, to abstract, and to ignore completely? This tension between parsimony and completeness will continue to haunt sciences of the artificial until hard evidence from the natural sciences can deliver either $e = mc^2$ for the brain or the bad news that *it takes a village* full of disparate mechanisms to produce truly general intelligence.

Of all the principles discussed in this book, predictive coding has the greatest potential of channeling neuroscience's inner Einstein.

6 Emergence of Predictive Networks

Emergence involves processes running at several timescales: evolutionary, developmental, and epigenetic (aka lifetime learning). Their interactions are so powerful and prevalent that viewing each in isolation can yield a faulty understanding and oversimplified computational models. Evolution produces genomes that serve not as blueprints for brain circuits but as recipes for brain growth and self-assembly during development, and one key determinant of the developing neuronal connective topology is the coincident firing of neurons on both ends of a synapse: the same process, spike-timing-dependent plasticity (STDP), that governs learning. Hence, it is hard to talk about brain evolution without invoking development, which overlaps with learning.

This chapter investigates the origins of predictive networks with respect to all three of these emergent levels. The weight of the discussion involves evolution, but development and learning get drawn in quite often.

The fossil record preserves skulls, not brains, so discovering the evolutionary history of neural systems demands the highest level of scientific sleuthing. Experts in this area have several tools at their disposal to help codify at least some of the key pieces of the puzzle. The old adage that *ontogeny recapitulates phylogeny*—as proposed in 1866 by Ernst Haeckel and thoroughly examined a century later by Stephen Jay Gould (1970)—provides ample justification for examining the embryology of modern animals in search of clues to the evolution of complex life, and intelligence.[1] Comparative anatomy and physiology of extant species also provide convincing evidence of transitions in cerebral structure and function across the millennia, since many of these species have inhabited the earth for many millions of years longer than humans. In addition, genetic analyses of those same species point toward particular mechanisms, such as genetically controlled timing patterns during development, that may underlie those transitions. These and other investigative tools are the basis of several excellent books on the evolution of brains (Allman 1999; Schneider 2014; Striedter 2005; Deacon 1998; Edelman 1992).

In *Intelligence Emerging* (Downing 2015), I weave together several of these accounts, starting with the replication with modification of Hox genes that produce ever more heterogeneous segmented morphologies, with brain segments multiplying and diversifying in the same way as body segments, leading to the *ice cream cone* model (Allman 1999) of the evolving brain. More formally, each scoop of ice cream corresponds to a *neuromere*, a ringed segment that functions as a modular zone of high cell division (producing neurons) and radial migration (of neurons to their proper layer) during brain development, as discovered by Bergquist and Kallen (1953) and further refined by Puelles and Rubenstein (1993) into what is officially known as the *neuromeric model* (Striedter 2005).

The stacked neuromeres gradually evolve into a hierarchical controller, with the spinal cord, medulla, cerebellum, and basal ganglia growing out of the lower part of the stack and handling many of the faster, subconscious sensorimotor activities, while the cortex and hippocampus sit atop the stack and handle slower, higher-level processing, including conscious reasoning and memory retrieval.

The view of brains as predictive machines does little to change these plausible evolutionary and developmental accounts of emergent neural circuits. However, it does entail a selective pressure for an ability to generate and evaluate expectations, and this interpretive bias would surely shift the anatomical and physiological focus in any of the noteworthy accounts cited above. This chapter takes that selective pressure for granted, so instead of providing a long list of examples in which different species gain a selective advantage via predictive machinery (i.e., the *why* of prediction), it focuses on the *how* and *what* regarding processes that have enabled anticipatory neural circuits to arise.

6.1   Facilitated Variation

Evolution involves three key processes: the production of genetic variation, the inheritance of genomes, and natural selection of the phenotypes produced by genomes. Kirschner and Gerhart's *theory of facilitated variation* (2005) addresses the first pillar, variation, that neither Darwin (selection) nor Mendel (inheritance) was fully equipped to explain. This goes well beyond DNA, crossover, inversion, and mutation, beyond the basic mechanisms of *how*, to the level of *how within the short span of a few billion years*. Because despite all of that time, the odds of purely random genetic change producing earth's levels of biological complexity are astoundingly long, even with the help of selection and inheritance. Kirschner and Gerhart have shown that the process is far from random and leverages an elegant hierarchy of emergent scaffolding.

The gist of facilitated variation is providing the grist for selection: over the course of evolution, genotypic variants become biased toward those with which selection can work to produce well-adapted populations. Thus, most genetic change is not lethal, and at least neutral; and the odds of producing design improvements become surmountable. This capacity to produce viable, inheritable grist for selection is termed *evolvability*, and it relies on two key characteristics: robustness and adaptability. Robust systems are those that can buffer the effects of external perturbations without creating much internal disturbance. For instance, in arctic mammals, a thick layer of subcutaneous fat provides a robust defense against extreme temperatures. Adaptability entails a more active role of internal mechanisms to combat external factors. For example, a dog adapts to hot weather by panting, to expel more internal heat; and from a longer time perspective, a canine species may evolve the ability to produce a thicker coat of hair if it gradually migrates to colder climates. In this sense, the species adapts (i.e., changes its genotype) to produce a more robust phenotype.

Facilitated variation involves three key interlinked mechanisms for attaining robustness and adaptability: modularity, exploratory growth, and functional coupling via *weak linkage*.

Together, these make a species robust to both environmental and genetic perturbations—most mutations are neutral—and adaptive, by making transitions to new, improved genotypes and/or phenotypes possible, if not highly probable. Thus, the canines that migrate to Siberia do not spend hundreds of generations shivering in the cold and producing small litters of weak pups. Instead, their genetic material and/or current phenotypes stand *poised and waiting* to handle many types of disturbances. Getting to that poised state requires modularity, exploratory growth, and weak linkage, combined and improved over long paths through the evolutionary tree.

Kirschner and Gerhart identify many *core processes* that exhibit or contribute to these three key properties. These have arisen at different points in the evolution of life on earth and are *conserved*, that is, remain vital to this day. The older inventions, aged three billion years or more, include energy metabolism, membrane formation, and DNA replication, while more recent biological breakthroughs (only a billion or less years old) are intercellular signaling pathways and body-axis formation (anterior-posterior, dorsal-ventral, and such).

Systems with high evolvability have reached a state in which most components have complex internal activity but reduced and simplified interactions with other components. This preponderance of internal over external interactions constitutes *modularity*, while the presence of many simple signals between components manifests *weak linkage*. This pertains to both phenotypes (in which organs are modular with weak linkage via chemicals in the blood stream) and genotypes, with modular substrings of base pairs comprising structural genes that produce complex protein molecules, but with regulatory signals that turn these genes on and off being relatively simple chemical compounds.

Modularity and weak linkage support a key corollary of facilitated variation: duplication and differentiation. Once the complexities of a unit are largely self-contained, with only weak connections to other units, the possibilities increase for copying it (with or without small variations) and successfully integrating the daughter unit(s) into the existing structure. This applies at both the phenotypic and genotypic levels. Neurons themselves, a phenotypic unit, can easily duplicate, with progeny cells easily linking into the growing network. Underlying the modular phenotypic body and brain segments (e.g., the ice-cream-cone brain model) are modular genetic components, such as the Hox genes, that have been duplicated and differentiated over the ages.

From an evolutionary perspective, the poised state is one in which the complex components are highly robust to genetic change, while the system as a whole achieves adaptability via a flexible network of weak linkages. The core processes only need to change the signals that they send or receive (via receptors) to alter the system's dynamic topology; and since these signals tend to be simple, abundant chemicals, such as calcium, no evolutionary time is wasted trying to invent them. In fact, the DNA contains many structural genes that already have the ability to create particular proteins but are inhibited by the nascent chemical network. Making a significant phenotypic jump from this state requires nothing more than a disinhibitor, for example, a chemical that binds to the active sites of the inhibitor. Weakly linked component topologies are easy to modify, but without messing with the finely tuned recipes for creating the complex components, recipes that took millions of years to evolve.

The third property, *exploratory growth*, typically refers to the ability of components to structurally connect to one another during development. For example, a neuron grows an axonal cone that follows chemical gradients in search of target dendrites. The type of

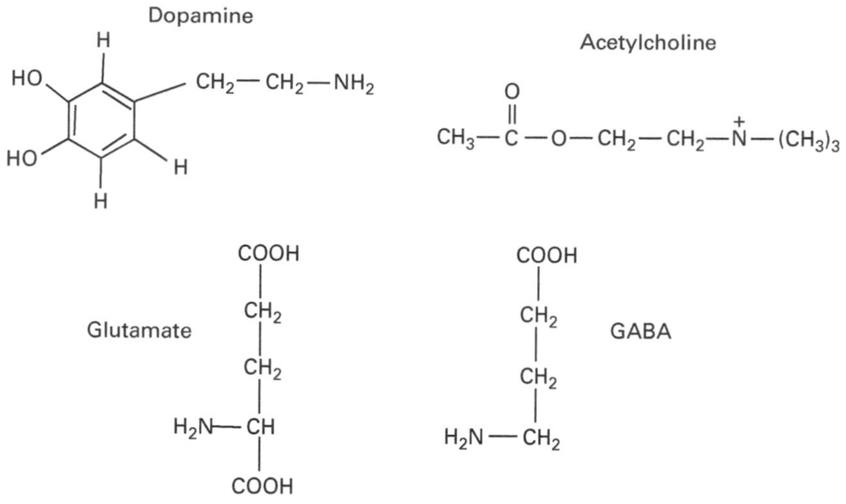

Dopamine

Figure 6.1
Molecular structure of four main neurotransmitters. The hexagonal ring in dopamine consists of six carbon atoms. Note that glutamate is simply CO_2 plus GABA (aka gamma-aminobutyric acid).

dendrite that it seeks may be encoded in the DNA, but its exact location is certainly not: the DNA encodes an algorithm or recipe, not a static blueprint. This ability to explore supports robustness in that one environmental (or genetic) change that alters the location of a developing module (M) in three-dimensional space does not require additional changes for partnering modules to find and connect to M; the exploratory growth process will eventually make its way to M and form the essential structural link. This also enables adaptability, since any *needed* change to a phenotype, such as a longer neck for giraffes, may require only a few genetic changes, with the rest of the body *growing to fit* the altered phenotypic trait. A longer neck does not require additional mutations for longer motor-nerve fibers when those nerves have already evolved the ability to grow to find their target muscles. Thus, the adaptive phenotypic change requires fewer genomic modifications and thus becomes a more accessible point on the evolutionary landscape.

Brains are classic examples of modular, weakly linked systems formed by exploratory growth, where the seven main neurotransmitters (all relatively small molecules) transfer signals across most synapses. In addition, the majority of interneuron signals involve either traveling depolarization waves or direct electrical stimulation (via gap junctions). Similarly, there are four or five main neuromodulators, all of which are very simple chemical compounds compared to proteins. In short, neurons communicate using a currency of only a few simple coins.

Figure 6.1 shows the chemical simplicity of four key neurotransmitters, three of which (dopamine, glutamate, and GABA) are amino acids or direct derivatives of them. In comparison, the simplest protein, ribonuclease, consists of 124 amino acids. Glutamate is the most common excitatory neurotransmitter in the brain, while the most prevalent inhibitor is GABA. Dopamine functions primarily as a neuromodulator; it is broadcast regionally by the basal ganglia. Acetylcholine (ACh) is the main transmitter employed by the spinal cord and is released at all vertebrate neuromuscular junctions (Kandel, Schwartz, and Jessell 2000).

The striking similarity between glutamate and GABA indicates that a neuron that emits one of them would require little genetic retooling to become an emitter of the other. Although the receptors for glutamate and GABA have different origins, most neurons contain receptors for both excitatory and inhibitory compounds. Thus, in nascent species, the key cellular adaptation that could significantly alter the overall neural signaling topology is a change in emitted neurotransmitter, with little or no change needed on the receiving end. The evolutionary emergence of predictive-coding circuits with their excitors, inhibitors, and comparators seems almost inevitable.

The simplicities and similarities of these neural signaling compounds, conserved across the animal kingdom, serve as a preliminary indicator of facilitated variation's role in brain evolution. This chapter digs deeper, in search of other indicators that predictive mechanisms were a natural corollary to emergent intelligence, not only because of their selective advantage, but also because of structural transitions enabled by a system poised and waiting for adaptive opportunities.

6.2   Origins of Sensorimotor Activity

As with many stories of evolutionary origins, the following is only speculative. If we view multicellular organisms as scaled-up versions of bacteria, for example, then their earliest forms should include both sensors and actuators, with the common view that all organisms sense first, then act according to the sensory inputs. However, this sense-and-act (aka outside-in) bias has been strongly contested by a host of prominent cognitive scientists (Buzsaki 2019; Clark 2016; Ballard 2015; Llinas 2001), many of whom view action as primary. Buzsaki (2019) argues that an organism that can only sense but not act would have no survival advantage, while an acting but non-sensing agent could still move (all or part of itself) and fortuitously acquire needed resources.

This argument is well worth noting in the context of prediction, since any system in which expectations play a major role can be one in which the dominance of sensory reality over internal goals or predictions comes into question. The inside-out view of cognition pervades much of the discussion in this and all remaining chapters, but we can nonetheless begin with primitive organisms that have documented abilities to both sense and act, with no necessary priority of one over the other in their cooperative synergy.

An early precursor module to nervous systems, as postulated by many neuroscientists, is a hybrid sensorimotor cell similar to that found in coelenterates such as hydra and sponges. As detailed by Schneider (2014), these cells detect external chemical and physical disturbances via cilia, which can then send electrical signals to a contractile region of the same cell. As shown in figure 6.2, when physically coupled together, the reactive behavior of individual cells can alter the shape of the colony in various selectively advantageous ways. For example, when one cell contracts, its neighbor cells might stretch so as to form a cavity.

In a subsequent evolutionary stage, the sensory and contractile regions might separate into two distinct cells that still communicate electrically, but now via a gap junction, as shown in figure 6.3 (upper left). Sensory cells could then congregate on the periphery, forming a dense ciliary mat, while muscle cells (aka myocytes) underneath would have an expanded sensory field via input from several sensory units. This represents a disjunctive contractile behavior: if *any* of the sensory units activate, contraction will follow.

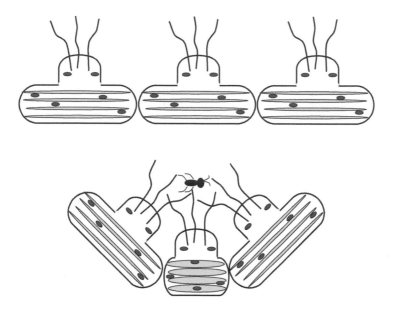

Figure 6.2
Caricature of early hybrid cells (based on those of contemporary coelenterates) that combine sensory (vertical hairs) and motor (horizontal strands) capabilities, as described by Mackie (1970). In response to sensory stimulation, the muscle fibers of the first-stimulated cell might contract, while follow-up inhibitory signals would prevent neighboring cells from contracting, thus creating invaginations to serve purposes such as capturing prey. Note that for larger prey, more cells would sense its presence, thus resulting in more widespread contraction and a larger feeding pocket.

Further spatial separation of sensors and myocytes permits a deeper embedding of motor activity in the organism's core, such that muscle contractions could cause more widespread physical deformation and/or regulate internal liquid or gas flows. However, this distance precludes direct, targeted electrical communication via gap junctions, thus requiring a *cable*, such as a neural filament (figure 6.3, upper right).

The introduction of intermediate neural cells (figure 6.3, lower right) ushers in a new era in which myocytes and sensors communicate only indirectly, via way stations that can accumulate signals from several sensors before stimulating a myocyte. This essentially adds conjunctive functionality to the system: certain myocytes may only contract when *several* sensors register a disturbance within the same short time window.

Finally, the proliferation of interneuron layers yields additional intricacy (i.e., more complex logic) to sensorimotor relationships (figure 6.3, lower left). Also, the emergence of chemical synapses permits filament behavior to change over time, implementing a primitive form of learning.

6.2.1 Origins of Oscillations

Oscillatory dynamics also appear to have evolved from external (muscle) to internal (cortical) structures, another classic example of *encephalization*, as defined by Llinas (2001): the incorporation of peripheral mechanisms into the nervous system. Basically, when cells housing contractile elements are strung together, the electrical coupling between each myocyte forms an active substrate capable of producing undulating waves, which can serve

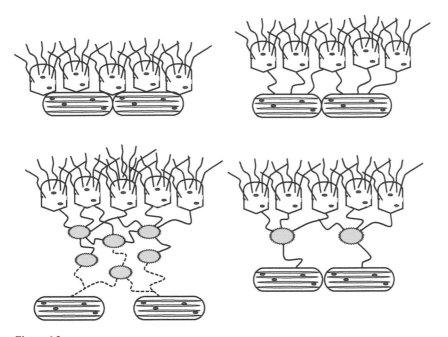

Figure 6.3
The evolutionary complexification of the neural sensorimotor network as envisioned by several neuroscientists (Llinas 2001; Schneider 2014; Mackie 1970). (Top left) The hybrid cell from figure 6.2 evolves into separate sensory and motor cells, coupled electrically via gap junctions. (Top right) Sensory and motor cells separate further and communicate along simple ionic channels with chemical synapses, but with gap junctions still possible between pairs of epithelial or contractile cells. (Bottom right) Interneurons arise as transitions between sensory and motor cells, with cable-like connections between all heterogeneous cells. (Bottom left) Proliferation of intermediate cells enables more intricate relationships between sensing and contracting, less reliance on gap junctions, and more adaptivity due to filaments containing modifiable synapses (dotted lines).

a wide variety of purposes: locomotion, digestion, excretion, circulation, and so on. Figure 6.4 illustrates this basic effect for sheets of coupled muscle fibers. The top diagram shows a simple traveling wave caused by sequential contractions, while the lower caricature of a heart muscle shows how a folded sheet of such fibers can act as a primitive pump.

The extant lamprey serves as a window into the early evolution of oscillations and their role in movement. As shown in figure 6.5, the lamprey's peripheral nervous system includes spinal-cord modules dedicated to body segments, with each module containing identical left and right neural groups. Each group receives input from a corresponding brain-stem area (via connections between B and E), and these inputs excite the motor neurons on that side while inhibiting all activity on the opposing side. Thus, contraction on the left forces relaxation on the right. However, this one signal may be enough to initiate a *fixed action pattern* (FAP) (in a particular segment) that requires no further reticular prodding.

After neuron E fires, thus stimulating M and its corresponding left-side myocyte, it also activates neuron I, which inhibits the opposite (right) side. However, E also excites the slower-charging neuron L, which eventually fires and inhibits the inhibitor, I. This disinhibition of the opposite (right) side initiates a phenomena known as *postinhibitory rebound*, wherein a strongly inhibited area depolarizes, despite the lack of excitatory input. Thus, the right side activates its corresponding motor neuron and myocyte, and inhibits the left side.

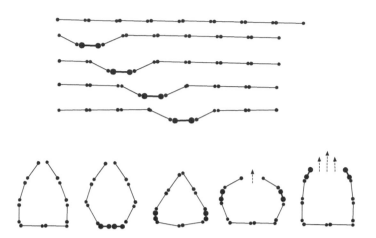

Figure 6.4
Activity waves in coupled muscle fibers (line segments), with thick (thin) segments representing contracted (relaxed) fibers. (Top) A moving invagination formed by sequential contractions. (Bottom) Pumping of liquid from a chamber via parallel chains of sequentially firing units.

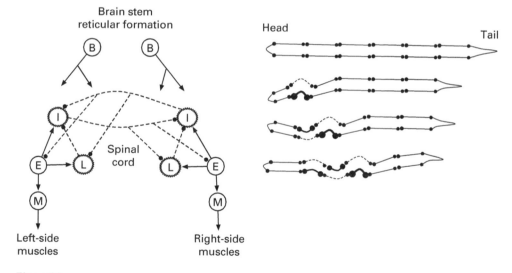

Figure 6.5
Swimming motion in the lamprey and its neural basis. (Left) Neural controller of a body segment. Solid lines are excitatory, while dashed lines are inhibitory. The four main neurons in each pair are I, fast distant inhibitor; L, slow local inhibitor; E, fast excitatory; M, motor. (Right) A wave of paired contraction (thick solid segments) and relaxation (dotted segments) combined with rebound contraction-relaxation produces slithering movement. See Kandel, Schwartz, and Jessell (2000, 745) for more detailed diagrams and descriptions.

When the right-side L neuron activates and blocks the right-side I neuron, a postinhibitory rebound occurs on the left side, and the cycle continues.

As portrayed in figure 6.5 (right), this cycle of paired excitation and inhibition produces the behavioral contraction-relaxation pair, with the former causing its side of the body to press inward and the latter producing an outward bend. Each pair connects to a neighboring pair (going from head to tail in the animal) such that activity in one pair spreads backward. As the main activity wave travels caudally (tailward) in the diagram, the oscillating rebound deformations move rostrally (headward), thus yielding the characteristic slithering motion of lampreys, eels, and snakes.

Notably, any pair of mutually inhibiting neural groups can form an oscillator if both groups exhibit postinhibitory rebound; and this cycle-producing module can reappear in higher brain regions in the course of evolution. Hence, a primitive motor mechanism, essential for many basic bodily functions, constitutes a *core process*, in the terminology of facilitated variation, which can then ascend the neural hierarchy to become the generator of an internal brain rhythm that may control motion as well as visceral and cognitive functions.

In what looks to be a textbook example of ontogeny recapitulating phylogeny, Llinas (2001) explains how direct electrical coupling between myocytes during early development of the shark produces *motor tremor*, which drives essential embryonic movement and fluid circulation in its prenatal environment. After several weeks, the myocytes lose their interconnections and become governed by electrically coupled motor neurons, thus ushering in the *neurogenic motricity* stage. The adult lamprey exhibits precisely this stage, with the neural controllers of each segment forming a chain of coupled oscillators, whereas in the adult shark, the oscillators have moved from the spinal cord up to the brain stem—and in mammals, they reside at all levels, from the brain stem, cerebellum, and basal ganglia all the way up to the hippocampus, thalamus, and neocortex. The earlier example (in chapter 3) of circuitry for recognition versus action generation illustrates how oscillators could operate in the premotor cortex.

Llinas (2001, 63) sums this up as follows:

How is this embedding of motricity actually accomplished and why should we want to know? Perhaps in understanding how this was/is accomplished, we will understand something about our very own nature, as the mechanism for internalization must be very closely related to how we process our own thinking, and to the nature of mind and of learning by experience. The short answer to the question, as we have discussed, is that we do this by activation and transfer of intrinsic, oscillatory electrical properties. Basically, motor neurons fire intrinsically, muscles contract rhythmically, and the receptors in muscles and joints respond to the movement and inform motor neurons about their success in producing movement and the direction of such movement in body coordinates. In other words, when I activate motor neurons, I get a sensory echo—and the echo somehow seems to be related to my body's response to the motor order.

The relationships between motor activity and this *sensory echo* is crucial to understanding the predictive brain. In its simplest form, the echo has a very limited range: from the muscles and joints to the spinal cord, where it contributes to local feedbacks that control body position. These spinal reflex loops provide an interesting variant of predictive coding, wherein expectations from higher cortical areas (e.g., motor cortex) function as goals for the body's mechanical subsystems. Roughly speaking, these goals are represented as desired (near future) sensory proprioceptive signals, such as those indicating muscular contraction, skin stretch, and joint rotation, as well as local pressures and temperatures.

Many cognitive scientists (Clark 2016; Shipp, Adams, and Friston 2013; Ballard 2015) draw attention to these spinal reflex loops as key evidence of predictive coding and *active inference* (Friston 2010) (i.e., modifying the sensory system to achieve states compatible with the mind's current expectations). Figure 6.6 illustrates the basic neural circuitry and regulatory process. Crucially, many of the body's motor control tasks require very little intervention from higher cortical areas, such as the premotor and motor cortices. They may

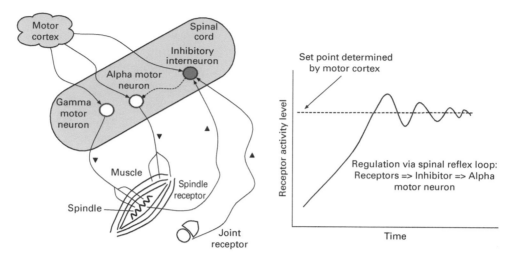

Figure 6.6
Basic mechanisms of muscle control. (Left) High-level sketch of corticospinal and spinomuscular relationships based on similar diagrams in (Kandel, Schwartz, and Jessell 2000) and discussions in (Clark 2016). (Right) Simple graph of the progression of activity in muscle receptors (in spindles, joints, and so on) toward a target level determined by signals initiated in the motor cortex and channeled through the gamma motor neurons, which adjust the spinal cord's sensitivity to the muscle receptors by effectively determining the gain / strength of the receptor pathways back to the spine.

send down a few initiating signals, but after that, feedback loops between the spinal cord (or other lower-level areas such as the medulla or cerebellum) and muscle units get the job done. The figure shows that signals from the motor cortex impinge on several spinal neurons—three of the main types are shown. These signals determine a baseline level of depolarization for each neuron, thus influencing how *jittery* or *relaxed* they will be, and this essentially determines the dynamic parameters and set point(s) for the regulator. For example, the firing rate of the gamma neuron affects physical properties of the spindle, which, in turn, determines the gain on the feedback from the spindle receptor to spinal inhibitory interneurons. These inhibitors form the key negative link in the feedback loop by adjusting activity of the alpha motor neurons, which directly control contraction of the muscle fiber.

The popular predictive-coding interpretation of this physiology is then straightforward. The higher cortical areas have predictions / goals, which they send to the spinal reflex systems, adjusting their parameters to implicitly form set points for the proprioceptive *reality* to be sampled. The negative feedback loops then ensure (under normal conditions) that sensory reality will be manipulated to match the goals, thus reducing the prediction error to zero. When the goal state consists primarily of proprioceptive readings, the concept of active inference seems wholly plausible, since an agent does have considerable control over its own body such that sensory reality can be molded to fit the goal.

As pointed out by Shipp and colleagues, the lower areas (i.e., closer to the spinal tract) of the motor cortex are poorly suited for receiving bottom-up signals from the spinal cord—due to a paucity of neurons in layer 4, the standard avenue of bottom-up projections in all cortical columns (Shipp, Adams, and Friston 2013). They also point out that these connections exist in the human fetus but largely disappear in infants. This indicates that the spinal system in

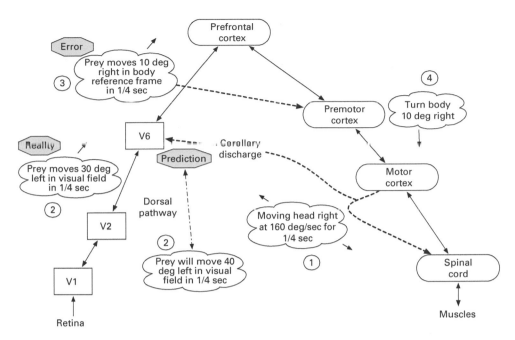

Figure 6.7
Illustrating the role of corollary discharge in predictive coding. V1, V2, and V6 are visual-processing areas. Circled numbers delineate the general sequence of activities. (1) The motor cortex signals the spinal cord to rotate the head 40 degrees in the next 1/4 second. That same information is sent to area V6 on the dorsal sensory pathway as the corollary discharge (aka efference copy). (2) This allows V6 to *predict* the prey's movement in the visual field, which it can then compare to the *sensed* movement, (3) with the difference being the prediction error, which also embodies the *perceived* movement, which is then sent back across lateral connections to the premotor cortex, (4) which computes a body rotation (to keep the prey centered in the body reference frame) and sends it to the motor cortex.

the fully formed organism has no need to send prediction errors upward in the hierarchy, since the low-level regulator will handle the problem (error), and the higher level has all the information that it needs: *the goal state is (or will soon be) achieved.*

When sensation goes beyond proprioception, to signals generated outside the body, the brain cannot assume such unmitigated success in achieving goal states, which now include the telereceptors, for example, those in the eyes and ears that detect signals at a distance. In this case, the prediction and sensory input may interact at higher levels, where telereceptor inputs get preprocessed.

The situation in figure 6.7 is one of the classic examples of predictive coding: the use of corollary discharge to form a prediction of upcoming sensory input, which is then compared to the (only partially preprocessed) sensory reality to yield a prediction error that constitutes the brain's actual perception. This basic process explains a wide range of human skills such as catching a flying object while running and rotating one's head, and oddities, for example, failing to tickle oneself. It also explains a bat's ability to differentiate self signals from those of potential prey during echolocation, and similar capabilities of fish that hunt using self-generated electric fields. Many optical and auditory illusions also stem from corollary discharge.

The contribution of oscillation in figure 6.7 is also important to note. Several of the (presumably neurally encoded) messages include the timestep, 1/4 second. Since cerebral signaling typically occurs in discrete chunks determined by the primary oscillations used between different areas at particular times, the timestep should be implicit. For example, if the motor cortex and V6 exchange signals on a 4 Hz cycle, then each velocity signal will implicitly entail *over the past 1/4 second*. If this is the standard frequency for interactions between these regions, then all neural circuitry in those regions will become calibrated for that fixed timestep, thus saving both signaling (sending of a timestep) and calculating (multiplying a timestep by a velocity) resources.

6.2.2   Activity Regulation in the Brain

Brain oscillations proliferate during embryonic development, even in the absence of external stimulation, as reviewed and computationally modeled by Abbott and coworkers (2003). These steady firing patterns are both regulated by and regulators of the growing organism in a complex self-organizing process. The rhythmic signal seems crucial for proper wiring of the neural circuitry and then persists into adulthood. Another seminal modeling study, by Ooyen and colleagues (2003), posits homeostatic mechanisms for maintaining activity set points by adjusting dendritic growth, thereby modifying the total input signal.

As another example of *weak linkage*, both studies highlight calcium ion Ca^{++} concentration as the main signaling factor, since it is typically positively correlated with neuronal excitation. In Abbott's group, which studied rhythmic activity in models of the digestive system of crustaceans, Ca^{++} deviations from a set point promote (a) ion-channel addition, removal, and alteration, and (b) changes of synaptic strength. Both regulatory actions stabilize Ca^{++} at the set point. Ooyen's team acknowledges the role of these ion-channel and synaptic mechanisms while examining more significant structural change as a factor in firing-level homeostasis. By allowing dendritic trees to grow and retract in response to Ca^{++}, their models attain stability in networks consisting of either all excitatory neurons or mixtures of excitatory and inhibitory units, with more intense dendritic growth in the mixed networks as some overinhibited neurons search for additional excitatory inputs. Their model includes the known principle that inhibitory neurons mature more slowly than excitatory cells, and this indirectly entails that they need not expand their dendritic tree when understimulated. The weak activation of an inhibitor stems, at least partially, from weak activity in afferent excitatory units, which adapt to increase their own activity by expanding dendritic scope; and this, in turn, increases stimulation to the inhibitors. Thus, the inhibitors need not expend a lot of energy growing dendrites if they can *be patient* and wait for neighboring promoters to do the job for them. Once again, self-organization works things out.

In general, these modeling studies (and considerable additional work cited by the two labs) indicates the ubiquitous control of neural firing rates beginning prenatally—in the simpler, preliminary versions of the brain—and continuing throughout life. Many other structural and functional properties of mature brains presumably stem from this homeostasis as well.

Predictive coding may be one of them. By comparing predictions to reality throughout the brain—not only at the output end—and working to reduce their differences wherever possible, the brain regulates the activity of many neurons, such as those representing prediction error, around low set points. Although the relationship between prediction and regulation and their evolutionary origins may seem like a chicken-and-egg problem, the selective advantage of energy minimization probably trumps that of prediction at the lowest levels of biological complexity. Thus, neural circuits that restricted neuronal activity levels and the density of neural group activity may have initially served the dual purposes of energy reduction and information maximization, with detailed thermodynamic and information-theoretic analyses (Stone 2018; Sterling and Laughlin 2015) pointing to metabolic efficiency as the absolute highest priority. Then, predictive coding could have emerged as an exaptation (i.e., alternate usage) of this regulatory machinery.

Peter Stone concludes his book on neural information theory with the following insights (Stone 2018, 166), which may apply equally well to predictive coding as to color opponency:

In the context of neural computation, the evidence collated from a wide variety of research papers suggests that a plausible candidate regarding why physiological mechanisms work in the particular way they do is because they maximize the amount of information gained for each Joule of energy expended (i.e., to maximize metabolic efficiency). In the process of testing this hypothesis, we have found that most functions (like color opponency) which were *once perceived as mysterious* [my emphasis] are actually implemented in ways that are *inevitable* [my emphasis], and even obvious, when considered *in the light of evolution*: specifically, in the context of neural information theory.

Regardless of primary, secondary, and tertiary purposes, the tendency for the brain to maintain optimal activity levels via both short-term dynamics and longer-term structural modifications plays a significant role in learning and development; these homeostatic processes begin in the early embryo and continue throughout life. By modifying genes that determine basic properties of ion channels, synapses, axons, and dendrites, evolution has gradually crafted set points and mechanisms for maintaining them. From all of this arises a complex neural controller that we can accurately call a *prediction machine*.

6.2.3 Competition and Cooperation in Brain Development

Neurons partake in an endless quest to influence neurons. Their own activity has little value if it cannot affect others. However, the brain, as a whole, cannot afford to have all neurons activating one another all the time, since such dense and widespread activity requires considerable energy and greatly reduces the information content of the resulting aggregate signal. Hence, inhibition is omnipresent in the brain: all regions have one or several types of inhibitory interneurons that arise during development by cellular duplication, differentiation, and migration, along with the (more abundant) excitatory neurons.

Inhibitory interneurons often create a competitive environment for neighboring excitatory neurons (e.g., granular or pyramidal cells), such that when one or a small group of these activates, it stimulates nearby inhibitors which then reduce the activity in other neighboring excitatory neurons. This produces *sparse codes* in terms of small groups of active neurons, which generally have higher information content while requiring less energy (Sterling and Laughlin 2015; Stone 2018). A small cluster of active neurons in one area can then transmit its sparse information signal to another area via excitatory connections; inhibitors tend to send signals only locally, so the active granulars or pyramidals constitute the transmitted *message*.

Basically, many neural areas are built for competition between nearby neurons for the right to cooperate with downstream neurons (by activating them). Conversely, there are times when nearby neurons must cooperate (i.e., fire together) in order to achieve downstream firing and thus transmit their message. Many of the brain's neurons have synapses from tens of thousands of other neurons and require hundreds or more inputs to provide enough post-synaptic stimulation to fire. Competition and cooperation need to achieve a delicate balance in the brain.

Nobel Laureate Gerald Edelman (1987, 1992, 2000) invokes neural competition as a central tenet in his theory of neuronal group selection (TNGS), also known as *neural Darwinism*. This is essentially a *survival of the best networkers*. During development, neurons migrate to their proper locations and then grow axons that search for dendritic targets. In order to survive, the neuron must form these connections and also achieve cooperative firing with downstream neurons. More neurons and connections are formed during development than ultimately survive in the adult brain; the losing (and thus disappearing) synapses are those experiencing few episodes of concurrent pre- and postsynaptic somatic activity, and the dying neurons are those with too few successful synapses. This pressure to hook up and cooperate is particularly intense during development, though a *use-it-or-lose-it* principle also applies to the mature brain.

Competition arises during development because of limited dendritic targets for the growing axons. If region R1 is much larger than region R2, and if both regions are genetically predisposed to target region R3, then R1 will tend to have a competitive advantage based on the sheer number of its axons. Even when some of R2's axons connect to R3, they will have less chance of cooperative firing, since R1 has more neurons and thus more chances for coincident firing of enough R1 neurons to produce activity in a given R3 neuron. All advantages go to the larger region: its enhanced chances of internal cooperation give it a leg up for external competition. Of course, any oscillations within a region that help synchronize internal firing can also provide a competitive advantage in the contest to innervate other regions. Hence, synchrony could help level the playing field if exhibited by R2 but not R1.

Terrence Deacon's displacement theory (1998) leverages neural Darwinism to show how interregional brain topology emerges from exploratory growth and competitive networking, with a minimum of genetic instruction. The DNA needs to encode only general timing constraints regarding the onset and duration for the neuron-generating processes of each region. The onset affects the context into which the new neurons arise, while the duration governs the size of each neural group. The context and size of a group then determine its competition and its relative competitiveness, respectively. The general result is that *large equals well-connected* and thus influential. Deacon (1998) sums this up as follows:

So although genetic tinkering may not go on in any significant degree at the connection-by-connection level, genetic biasing at the level of whole populations of cells can result in reliable shifts in connection patterns. ... Relative increases in certain neuron populations will tend to translate into the more effective recruitment of afferent and efferent connections in the competition for axons and synapses. So, a genetic variation that increases or decreases the relative sizes of competing source populations of growing axons will tend to *displace* [my emphasis] or divert connections from the smaller to favor persistence of connections from the larger.

Related work by Finlay and Darlington (1995) promotes the adage that *late equals large*: regions that form later in development tend to be larger. Joaquin Fuster's (2003) seminal research on the cortex and its development supports this principle, since, in primates, the posterior sensory cortices and intermediate motor cortices mature sooner than the frontal cortical areas, which are larger and play a dominant role in monkey and human perception and motor control.

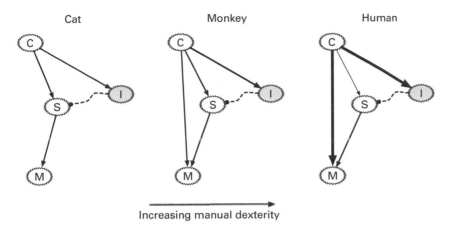

Figure 6.8
Gradual invasion of cortical regions into the motor neurons of the lower spinal cord across mammalian species based on text and diagrams in (Striedter 2005). C = excitatory cortical neurons, S = excitatory spinal neurons, I = inhibitory spinal neurons, M = motor neuron. Arrow thickness denotes strength of influence. Solid arrows are excitatory, dotted links inhibitory.

Striedter (2005) combines these results to show how small changes in developmental timing have led to later maturation in frontal regions, which enables them to grow larger and achieve greater control over a large expanse of lower cortical areas, which, in turn, seems to enhance behavioral sophistication. To wit, in studies of higher mammals, Nakajima et al. (2000) show a strong correlation between manual dexterity and the degree to which frontal cortices control motor areas.

Details of the neural circuitry investigated by Nakajima and coworkers reveal interesting parallels to predictive coding. Figure 6.8 sketches the general progression from cats to monkeys to humans in terms of the amount of cortical control of motor neurons. In the cat, cortical outputs primarily influence the upper spinal cord, which then controls the muscles via the motor neurons (in the lower spinal cord). Thus, most motor actions involve fixed-action patterns (FAPs) produced by central pattern generators (CPGs) in the the spinal circuits; the cortex merely sends enabling and disabling signals to the CPGs. In monkeys, the cortex exerts some direct control on the motor neurons while exhibiting more control over the inhibitory spinal neurons. Finally, in humans, motor neurons experience strong control from the cortex and comparatively less from the spinal cord. In addition, the cortex exerts a strong inhibitory effect on the excitatory spinal neurons.

This work highlights a competition between the cortex and spinal cord for motor control. It was originally believed that the cortex won that conflict via Deacon's displacement: as the cortex expanded from cats to primates, it simply out-fought the spinal cord for motor targets. However, Nakajima's lab found that the spinal-motor connections still exist but that the cortex increases its stimulation of spinal inhibition as a supplementary competitive *tactic*. Hence, the inhibition of one layer by another might simply emerge as an evolutionary consequence of the interregional battle for postsynaptic targets.

Combining this interlayer competitive inhibition with the previous principles of late-equals-large and caudal-to-rostral (back-to-front) cortical maturation yields one plausible account of the evolutionary and developmental (aka *EvoDevo*) emergence of cortical hierarchies configured for predictive coding. If the lower (caudal) cortical areas mature earlier than higher-level (rostral) regions, then the latter should contain more neurons and thus have a competitive advantage for innervating targets. And the greater that advantage, the more excess connections available to inhibit the lower-level competitors. Thus, a natural relationship between a higher and lower layer could be one of inhibition, and one easily achieved by tapping into the inhibitory interneurons already established in each region (as opposed to sending many individual inhibitory connections from the higher to lower level).

As further support for the ontological emergence of predictive-coding circuitry, consider the perspective of a single neuron (X) sending out axons. Under neural Darwinism, X needs to connect to targets and simultaneously activate with them. However, if X directly excites too many targets, then that will (a) provide a lifeline to neurons that can then compete with X for other targets, and (b) produce an over-stimulated network. One way to balance these trade-offs is to stimulate inhibitory neurons. These can co-fire with X, thereby keeping X well-connected and thus alive, while simultaneously depressing many of X's potential competitors and preventing an overheated network. In short, by exciting inhibitors, a neuron enhances its own survival while also helping to modulate overall network activity (and thus enhancing the survival of the organism as a whole).

An important detail omitted from our earlier discussion of cortical columns is that the axons of layer-5 pyramidal neurons realize both top-down links to lower columns and excitatory connections to motor neurons in the spinal cord and, in general, other motor areas, such as the superior colliculus, which controls eye movement (Mountcastle 1998; Hawkins 2004). Thus, one key function of cortical columns (at least in primates) involves motor control. Contrary to the view of the brain as a strict sense-and-act hierarchy—where all sensory signals travel upward through sensory cortex and then premotor signals travel back down along the *motor side*—these layer-5 motor projections appear in the cortical columns of all regions, whether primarily sensory or motor.

The layer-5 projections to other, nonmotor, columns may emerge as by-products of competition for the motor targets during development, especially if we accept the view that top-down links have a net inhibitory effect by exciting local interneurons in the lower columns, either directly or via more indirect means, as described earlier in Hawkins and Ahmad's (2016) model: initial, weak, top-down excitation leads to more widespread inhibition when predictive signals properly anticipate sensory reality.

This view of layer-5 outputs as general-purpose *action* signals—whether promoting or blocking physical movement or prediction—meshes with the view of each higher level functioning as a controller of its lower neighbor(s). Each regulator outputs a controlling action designed to reduce the prediction-error signal.

Neural selectionism versus constructivism
Edelman's TNGS essentially views synapses like fish eggs: many are produced, but only a few survive. This *selectionist* view has been contrasted with neural constructivism from Quartz and Sejnowski's (1997) frequently cited "Constructivist Manifesto." Constructivism views development and cognition as highly intertwined mechanisms, with significant amounts of ontological growth driven by demands of the brain-body-environment interaction. It weighs in on the side of *nurture* in the classic nature-nurture debate by emphasizing flexible neural structures molded by experience over brain topologies built primarily by

innate recipes in the DNA. In so doing, it further blurs the divisions between development and learning by proposing that both (a) the former extends well into life, and (b) the latter plays a vital role in the former. They see this flexibility as an evolutionary trend that increases in more behaviorally complex lifeforms. In their words (Quartz and Sejnowski 1997, 537),

Neural constructivism suggests that the evolutionary emergence of neocortex in mammals is a progression toward more flexible representational structures, in contrast to the popular view of cortical evolution as an increase in innate, specialized circuits. Human cortical postnatal development is also more extensive and protracted than generally supposed, suggesting that cortex has evolved so as to maximize the capacity of environmental structure to shape its structure and function through constructive learning.

The differences between selectionism and constructivism were greatly exaggerated by early critics of the manifesto such that a fictive dichotomy arose wherein either (a) the brain produces nearly all of its synapses prenatally, or neonatally, and then gradually prunes them away (in direct or indirect response to environmental signals) during childhood, adolescence, and adulthood, or (b) the brain generates axons, dendrites, and synapses in response to environmental signaling and internal neural dynamics throughout life, though primarily during the earlier decades (for humans).

A closer reading of the paper clearly indicates more of a compromise: constructivism posits adaptive growth *and elimination* of neural processes throughout life as part of the dynamic interplay between brain, body, and environment. Experimental neuroscience has since provided strong evidence for Quartz and Sejnowski's general claim (Petanjek et al. 2011; Sanes, Reh, and Harris 2006; Andersen et al. 2007), with significant structural adaptations (i.e., neurogenesis along with axonal growth and death) occurring well into the third decade of life.

In another seminal book, *Natural-Born Cyborgs*, Andy Clark addresses constructivism (Clark 2003, 84):

The human neocortex and prefrontal cortex, along with the extended developmental period of human childhood, allows the contemporary environment an opportunity to partially redesign aspects of our basic neural hardware itself. The designer environments ... are thus matched, step-by-step, by dedicated designer brains, with each side of the co-adaptive equation growing, changing and evolving to better fit—and maximally exploit—the other.

This view extends the cooperative-competitive narrative from the internal, interneuron level to the bigger picture of the dynamic, brain-body-world coupling. As discussed in chapter 7, this has strong implications for biologically inspired pursuits of artificial intelligence.

6.2.4   Layers and Modules

The brain is often portrayed as a hierarchical controller that continuously loops through sense-and-act cycles, wherein raw external inputs receive preprocessing on their way up the sensory hierarchy; then, the final, veridical perceptions at the higher levels initiate a signal sequence down the action side of the hierarchy, eventually producing movement. Though this workflow surely occurs in some situations, particularly those involving conscious thought and action planning, the neuroscience reveals an alternate picture (Mountcastle

1998). First, the hierarchy is more of a heterarchy: though many neural layers reside closer to the body-world interfaces (sensors and actuators) and may be in an exclusive group that exchanges signals with some section of the periphery, there are many such low-level layers that send lateral axons to the *other side* and that receive considerable stimulation from high-level layers. In fact, the often-cited ratio of top-down to bottom-up wiring in the brain is 10:1 in favor of top-down. Basically, the static architecture and functional dynamics of the brain indicate a tangled web but with undisputable evidence of heterarchy. Hence, any attempts to cleanly match predictive-coding modules to particular brain regions or cortical columns within the neural web will surely fail.

As mentioned at the outset, many prominent scientists (Buzsaki 2019; Clark 2016; Ballard 2015; Llinas 2001) prefer an inside-out, act-and-sense interpretation of the brain (over the traditional outside-in, sense-and-act view), which profits from generating models of the world that enhance the organism's ability to act, but not from capturing extraneous details of no obvious survival advantage. There is no inherent gain in building accurate mental models, nor in generating precise forecasts. Value comes only in the service of action.

Andy Clark eloquently sums up this perspective as follows (Clark 2016, 250):

In fact, these distinctions (between perception, cognition and action) now seem to obscure, rather than illuminate, the true flow of effect. In a certain sense, the brain is revealed not as (primarily) an engine of reason or quiet deliberation, but as an organ for the environmentally situated control of action. Cheap, fast, world-exploiting action, rather than the pursuit of truth, optimality, or deductive inference, is now the key organizing principle.

As for prediction's pivotal contribution to this situated control, he adds (Clark 2016, 250),

What real-world prediction is all about is the selection and control of world-engaging action. Insofar as such agents do try to "guess the world" that guessing is always and everywhere inflected in ways apt to support cycles of action and intervention. At the most fundamental level, this is simply because the whole apparatus (of prediction-based processing) exists only in order to help animals achieve their goals while avoiding fatally surprising encounters with the world.

This interpretation of the brain as a tightly integrated, act-and-sense heterarchy complicates our quest to neatly enumerate the core modules and weak linkages that could paint the predictive mind as an easily achieved variation, given a few hundred million years of evolution. Although cortical layers and distinct sensory and motor pathways look nice on paper, the microscope and MRI seem to tell a different story, both architecturally and functionally. However, within that confusing network reside hints of a predictive coder's building blocks.

As shown above, the interactions among motor cortex, spinal motor neurons, and muscles portray a simple controller whose prediction / goal, provided by the motor cortex, seems readily attainable by a reflexive arc. At higher levels, when corollary discharge forms the basis of a sensory prediction, the resulting prediction error constitutes the actual perception that the brain may use for conscious reasoning and/or further iterations through an act-and-sense loop. In both cases, the predictive-coding module spans several layers of neural circuitry, and at higher levels, the scope can easily include both sensory and motor regions. Whether the actual predictions begin as action requests or expected causes of sensory data (during a pure object-recognition task), they would appear to ground out in sensory expectations, which can then be compared to proprioceptors, telereceptors, or partially preprocessed signals in the thalamus or lower cortical areas. The central component in all of these cases is some form of comparator, which easily arises in a network replete with both excitatory and inhibitory neurons.

As the story of corollary discharge seems to indicate, prediction error serves as a perception, which, as such, consists of a sensory *reality* term minus a prediction. This paints perceptions as some form of abstraction, as opposed to concrete facts about the agent-world coupling. Abstracted representations are those with many details removed, so in this case, those details are the predicted components, the expectations. Notably, the perception does, in many cases, represent a more objective, ecocentric interpretation of the world. For example, the vestibular-ocular reflex gives the true picture of the world: it is not moving even though the head is. But from an egocentric, action-oriented view of life, which organisms have presumably evolved to honor, the *subjective truth of the matter* is that the world is moving along with the head. Still, the abstraction, a more stable world (the prediction error), is the representation needed to support the ensuing action, for example, stretching and snapping the frog's tongue to catch a fly.

In short, subjective reality (as recorded by an organism's sensory apparatus) constitutes the primary world view, with the maximum amount of detail (of any relevance whatsoever to that species). Predictive coding then filters and normalizes that information (relative to predictions) to create abstractions that more faithfully and efficiently support further action. In some cases, those abstractions provide more objective pictures of the world, while in others they are *helpful illusions* that aid survival (at least on average). The widespread rustling in the bushes may be a mountain lion, or a scurry of chipmunks, but better to be safe than sorry.

Thus, what begins at the periphery as relatively standard control—with the goal of reducing a prediction error defined as the difference between expected and experienced proprioception—becomes a regulator of more abstract information in moving up the neural heterarchy. From an action-oriented perspective, the brain is quite content if most abstractions wither away to nothingness on the way up the ladder, indicating an absence of surprise; and the activity can continue, as *planned*. Conversely, any unexpected sensory readings can create far-reaching abstractions that trigger action changes (and ensuing prediction updates) deeper in the nervous system.

As shown in figure 6.9, the neural heterarchy can, as a first pass, be modeled as a predictive-coding hierarchy of standard PID controllers, where each level works with different abstractions. Upward signaling involves the error terms (e_i), while downward messaging maps a *control variable* (u_i) to a prediction (p_{i-1}) via a transformation matrix of weights (M_{i-1}). At each level, the PID controller (C_i) converts the error (e_{i-1}) into u_i using the combination of the proportional (P), integral (I) and derivative (D) term, each sketched in the inset of figure 6.9.

At the lowest level, sensory input (s_0, predominantly from proprioceptors) is compared to target / predicted settings, producing an error sent upward to controller C_1, which produces a control response u_1, which, in turn, maps downward to possible updates of the expected result p_0 and gain g_0, which represents the attention paid to e_0.

The control decision, u_1, then becomes an aspect of *reality* to be compared to a higher-level prediction, p_1, derived from a higher-level control decision, u_2 (based on $e_1 g_1$), and so on up the hierarchy. At each level, the timescales vary, with higher-frequency updates (i.e., smaller time constants) at the lower levels. In this interpretation, the higher level tries

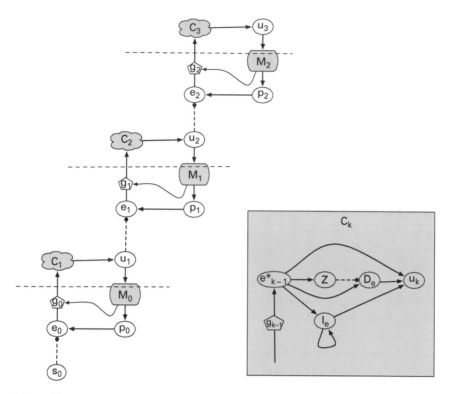

Figure 6.9
Predictive coding as a hierarchical controller, from the action-oriented perspective, where predictions constitute targets. (Left) Dotted horizontal lines separate layers according to conventional predictive-coding models. s_0 = sensory input, e_i = error, p_i = prediction, u_i = control variable produced by controller C_i, M_i = a model, a collection of weights / synapses mapping u_{i+1} to p_i, and g_i = gain applied to bottom-up signals. (Right, inset) Details of a neural PID (proportional, integral, derivative) controller circuit, with output u_k a weighted combination of the (scaled by g_{k-1}) error (from the layer below), the derivative of that error, D_e, and the error's integral I_e; neuron Z achieves the delay for computing D_e.

to predict only what its lower neighbor is *doing* (u_1), not what information it used to *decide* on that action.

From an alternate perspective, one more compatible with an act-and-sense view, p_1 indicates to level 1 what level 2 plans to do. If bottom-up information ($e_0 g_0$) presents a picture of reality that induces C_1 to produce an action (u_1) similar to p_1, then the ensuing low error (e_1) tells layer 2 that its action can be carried out, as planned. Conversely, a high error forces layer 2 to either modify its intentions or reduce its attention to e_1, by lowering g_1. In short, the higher level is telling the lower level its plans and then trying, via feedback, to reconcile them with the bottom-up reality signals—by either changing its plan or reducing the importance of the reality signal.

The primitive building blocks of the hierarchical controller also suffice to configure a value system similar to the basal ganglia (discussed in chapter 3). Figure 6.10 depicts the interplay between a predictive-coding hierarchy and a value system, which also receives control output u_3; but instead of transforming it into a prediction of a lower-level control output, it maps u_3 to an evaluation (V), thus mimicking the *critic* of a reinforcement learning (RL) system. In RL systems, these evaluations constitute predictions of cumulative future reward.

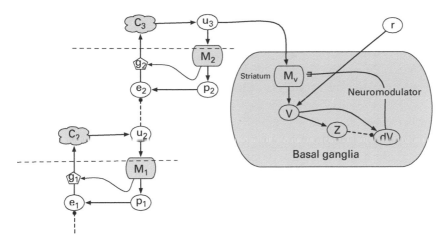

Figure 6.10
Enhancement of figure 6.9 to include reinforcement learning via an auxiliary network similar to the basal ganglia, where model M_V functions similar to the striatum and neighboring areas by mapping states to evaluations (V), while r denotes a reinforcing reward or penalty. The inverted delay connection between V and dV computes the temporal derivative of V, the *temporal difference error* described in chapter 3, which then governs the secretion of neuromodulator into the striatum, thus stimulating synaptic modification.

Employing a standard delayed inversion, the circuit then computes the temporal gradient of that prediction (dV), which represents RL's temporal-difference error (aka surprise). Neuromodulation (via chemicals such as dopamine) then responds to surprise, causing learning via synaptic change in the striatum and neighboring areas.

The similarity of building blocks between the hierarchical layers and the value system indicate the tractability of evolving either of them once nature had discovered the other, or the potential for coevolving these structures. The most significant difference lies in the neuromodulatory mechanism, which involves diffuse chemical signaling as opposed to targeted synaptic transmission. Yet, several common neuromodulators, such as dopamine and acetylcholine, also function as neurotransmitters, thus easing any evolutionary transitions from a signal-transmitting to a signal-spreading region (or vice versa).[2]

6.2.5   Running through the Woods on an Icy Evening

As a concrete, though clearly speculative, example of this heterarchical controller in action, consider the task of running on an icy trail. At the lowest level, the leg/feet joint and muscle proprioceptors frequently deviate from their expected values (p_0) due to omnipresent slippage and simple, high-frequency, reflex responses. Thus, e_0 cannot be completely ameliorated, and its scaled (by g_0) and accumulated (by I_e in C_1) value translates into the need for a higher-level adjustment (u_1), such as a greater angle of foot contact (landing on the heel and thus creating a higher pressure point to dig into the ice-coated dirt). However, this altered landing angle conflicts with the predicted angle (p_1) and this error (e_1) will, with sufficient gain (g_2) trigger C_2.

The predicted landing angle (p_1) derives from u_2, which represents a general concept of *stride*, a combination of step distance and body elevation (among other things). For example, if stride consists of a long step distance but a low elevation, the proper (predicted) landing angle will probably be high, whereas increased height facilitates a flat-footed or toe landing.

Thus, changes to stride (u_2) can modify the predicted landing angle (p_1) to better match that chosen by C_1. On a mildly icy trail, digging in with the heels works fine, but when surface friction decreases and hardness increases, the better response is to land flat-footed for maximum surface contact. Either way, the change to landing angle will affect the stride, and vice versa. Once again, the scaled and accumulated landing-angle error will cause adjustments (via controller C_2) to the stride. Signal flow back and forth between these two lower levels will continuously regulate the running motion.

Now throw an obstacle into the mix: rounding a sharp bend, a large branch comes into view, five meters ahead. The telereceptive (visual) system notices this surprise, after several levels of processing, and feeds that information to motor control at a high level, for example, C_3, whose output (u_3) represents a particular stride sequence, for example, a combination of bounding and stutter steps to set up the proper foot plant that immediately precedes the hurdling motion. Thus, u_3 translates into predicted / goal strides, which must broker some sort of compromise in terms of landing angles so as to avoid extreme slippage while properly preparing for hurdling (which is difficult from a heel-first landing and normally requires some pushoff with the toes).

These continuous top-down, bottom-up compromises cloud the issue of whether behavior consists of iterated sense-and-act or act-and-sense sequences, but this only reflects the bigger picture of the body and environment engaged in an interactive loop with frequently murky distinctions between causes and effects. Hence, the inputs to each error unit (p_i and u_i) may never neatly represent a (relatively fixed) target (e.g., a set point) and a more malleable factor. Instead, the complex dynamics of the circuitry may force one variable to conform to the other in one situation while yielding a reversal of that dominance in another scenario.

Likewise, the representation of each error node as the sum of a positive and negative term is a drastic over-simplification, but if predictive coding is a viable theory of cognition, the standard output of an error region should be lower when predictions align with subjective sensory reality. Which term, the top-down prediction or the bottom-up sensory flow, is the more negative or positive in that interaction might vary among neural regions and still capture the essence of predictive coding. In general, inhibition is everywhere in the brain, so the opportunities for one set of signals to cancel out another are ubiquitous.

The point of this example, and of figure 6.9, is hardly to provide the structural and functional neural architecture of action, but merely to sketch a general heterarchy of feedback loops that should apply equally well to neural firing patterns that represent abstract concepts (e.g., the approach-to-hurdle stride sequence) as to those signaling an over-rotated joint. And each can, through a proper model, M_i, translate into firing patterns at another level that represent a different abstraction (e.g., stride or landing angle) in a different coordinate system.

In addition, the modularity of this type of system would seem to facilitate evolution in the same way that Rodney Brooks' subsumption architecture (1999) for robotic control supports a gradual enhancement of physical dexterity via behavior-unit stacking. In figure 6.9, the horizontal lines carve up the system in accordance with predictive coding, where the errors of one level feed into the next layer, which uses those errors to help compute the control value, which it then sends back down in the form of a prediction and an error gain (as determined by the transformational model, M_i). The error gain manifests *attention* in the predictive-coding framework, as emphasized by Andy Clark (2016, 59):

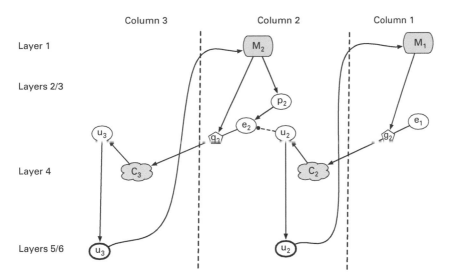

Figure 6.11
Hypothesized overlay of the predictive-coding topology of figure 6.9 onto adjacent cortical columns, separated by dotted vertical lines, with higher-level columns to the left. Transformational models (M_i) map top-down signals to predictions and gains via the highly adaptable synapses of layer 1. Layer 4 predominantly receives strong bottom-up signals, while layers 2 and 3 house a wide variety of excitatory and inhibitory neurons that could implement the basic PID controller (figure 6.9, inset). Layers 5 and 6 may amplify (thick circles) the control signal (u_i) from layer 2/3 before sending it to a lower level.

Attention, if these stories are on track, names the means or process by which an organism increases the gain (the weighting, hence the forward-flowing impact) on those prediction error units estimated to provide the most reliable sensory information relative to some current task, threat or opportunity.... The general idea is thus that patterns of neuronal activation (among the so-called "representation units") encode systematic guesses concerning task-relevant states of the world, while changes in the gain (i.e., changes in the weighting or "volume") on the associated error units reflect the brain's best estimate of the relative precision of the top-down "guessing" and the bottom-up sensory information.

This model maps relatively neatly onto the neural architecture of cortical columns, as shown in figure 6.11. As predictive-coding dictates, the main bottom-up signal is prediction error, and cortical models typically channel such information from layer 2/3 of the lower column to layer 4 of the higher column (Mountcastle 1998; Kandel, Schwartz, and Jessell 2000). Within layer 4, and particularly layers 2 and 3, these inputs can stimulate considerable computation that can also incorporate top-down predictive signals. Thus, layers 2–4 are likely candidates for the neural circuitry of the PID controller, including the computation of the layer's error signal. The control variable, u_i, is needed for both error computation and top-down signaling, so its value may be transferred to the large pyramidals in layers 5 and 6, which can strengthen the signal before sending it to layer 1 of a lower column.

It appears that predictive coding can piggyback off of the modular nature of cortical columns in order to satisfy a few basic requirements for evolvability: modular components with weak linkage between them. To scale up a brain composed of many such modules, a genetic mutation or two could probably generate extra columns, which could then hook to earlier columns using the same, well-tuned, developmental recipe for exploratory axonal growth. Thus, the extra columns would begin life with bottom-up synapses entering layer 4, top-down axons exiting layers 5 and 6, and top-down synapses entering in the dendritic mat of layer 1.

At each incremental stage of evolution, the uppermost cortical columns house controllers whose output values serve as the basis for both a downward prediction and a local error term. The control value and error provide *hooks* for any newly evolved higher layer to latch onto by creating a precision weight (gain, g_i) on the error and an expectation (p_i) for the control, with both determined by a model, whose synaptic weights presumably adapt as part of the learning process. Thus, each new layer arises as a meta-level above the last: it functions to modulate the gain of and predict the value of the lower layer's main products, e_i and u_i, using an even coarser level of abstraction (relative to the subjective sensory reality).

Three main tools of facilitated variation would thus seem to support the gradual evolution of a layered (hierarchical or heterarchical) predictive brain: modular core components (the cortical columns), weak linkage (standardized synapses such as those containing many NMDA receptors in layer 1), and exploratory growth (during both neural migration and axonal extension). The hardest part of adding additional functional modules would appear to be the crafting of each new transformational model, M_i, since these would have micro-circuitry and synaptic strengths tailored to the representations of the juxtaposed columns. These might have varying time windows of flexibility (aka critical periods), with lower-level models becoming fixed early in life, while higher-level models would maintain plasticity longer or indefinitely, as the neuroscience literature indicates (Sanes, Reh, and Harris 2006; Bear, Conners, and Paradiso 2001).

6.2.6   Oscillations and Learning

Assuming that the modular representation of cortical columns facilitates variation and thus enables evolution to scale up the brain by adding extra columns, which a well-refined developmental process (replete with exploratory growth) can easily interconnect, the final adaptive chore in building a predictive-coding network is the fine-tuning of synapses, a process that surely requires experience in the world. The architecture of cortical columns combined with two of the brain's most fundamental activities, oscillation and spike-time-dependent plasticity (STDP), provide interesting hints as to how this might happen.

In an attempt to explain learning of a hierarchy of abstractions, where higher-level abstractions predict those at lower levels, this section delves a bit deeper into structural and behavioral details of cortical columns. As shown in figure 6.12, the signal flow between adjacent cortical columns and within individual columns is not symmetric. There are well-established directional patterns that, although not flawlessly copied (as in an engineering device), do appear with surprising regularity across both brain regions and primate (and even many mammalian) species (Mountcastle 1998; Thomson and Bannister 2003). As described earlier, the main bottom-up flow goes from layers 2/3 in column C to layer 4 in column C + 1, then on to layers 2/3 in C + 1 and to layer 4 of column C + 2. In addition, layer 3 sends considerable excitation to layer 5 (and some to layer 6). Layers 5/6 feedback to layer 4 while also initiating the column's top-down response: those in column C send excitatory signals to the layer-1 dendritic mat of column C − 1, which has a strong modulatory effect on the activity of the entire column, but particularly layers 2/3. In short, the bottom-up feedforward path runs through layers 2–4, with the layer 3 axons of column C synapsing

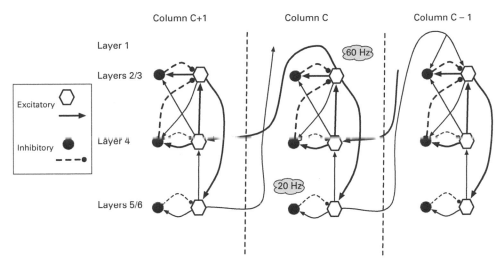

Figure 6.12
Main connections in and between cortical columns, based on diagrams and descriptions in (Bastos et al. 2012; Thomson and Bannister 2003). Each shape (hexagon or ragged circle) denotes a population of thousands of neurons; thicker (thinner) lines denote denser (sparser) connections. Many intercolumnar links are missing or partial to avoid clutter.

proximally onto column $C+1$'s layer-4 neurons, thus driving their activity in a relatively dominant manner. Conversely, the top-down feedback route involves the lower layers, 5 and 6, whose synapses into column $C-1$ tend to be distal, and thus less dominant and more modulatory.

Several additional details in figure 6.12 deserve special attention. First, although layer 4 does house inhibitory interneurons, their effects seem minor compared to the excitatory contributions from layer 3 of the downstream column, and from its own column's layer 3 (via layer 6—shown as layer 5/6 in the diagram). Layer 4 receives no significant inhibition from any other layer. Layer 2/3 neurons display an opposite connectivity pattern: they only receive strong excitatory input from one area, their own layer 4, but exhibit dense intra- and interlayer inhibition. All of this indicates that layer 4 in column C would tend to amplify signals from layers 2/3 in column $C-1$, whereas layers 2/3 in column C would sparsify the patterns feed in from layer 4, with that conversion modulated by top-down inputs from layers 5/6 of column $C+1$.

Another important, though often overlooked, characteristic of cortical columns involves interlayer differences in oscillation frequencies. As emphasized by Bastos and colleagues (2012), neurons in layers 2/3 have resonant frequencies in the gamma range, with dominant values around 50–70 Hz, whereas layers 5/6 tend to oscillate more slowly, in the beta range: 20–40 Hz. The lower layers appear to integrate signals from layer 3 (over several cycles) before firing and sending their top-down messages. The higher frequencies in layers 2/3 match the inhibitor-prevalent cytoarchitectural profile, since GABA receptors, the main channels for inhibition, have typical decay times in the 10–25 millisecond range, thus inducing oscillations in the 40–100 Hz range (Buzsaki 2006).

This beta-gamma distinction may exemplify the *callup mechanism* hypothesized by Buzsaki (2006, 2019), wherein signaling between two areas begins with the eventual receiver (Re) sending a low-frequency oscillation as a *request* to the sender (Se), which then resonates at a higher frequency. Thus, Re receives multiple waves of input from Se during each of its lower-frequency cycles: fast waves riding atop slow waves. Thus, in one of its normal cycles, Re integrates several signals from Se to create a compressed, that is, more abstract, spike train. Note how this also dovetails nicely with Buzsaki's inside-out view of the brain, since the higher (inner) areas (columns) initiate message-passing by calling up information from lower (outer) columns.

Another piece to this puzzle involves the competitive nature of neural layers containing considerable inhibition. When one neural assembly (A) activates, it can strongly block another assembly (B). However, when that depression wanes, the neurons in A may have habituated and/or those in B will exhibit a form of postinhibitory rebound (described earlier), thus giving B the upper leg in the next round of competition. Also, as discussed earlier, active cell assemblies strongly inhibit neighboring neurons and themselves for 100–150 milliseconds, thus giving neurons in other areas a chance to dominate. In this way, a high-frequency oscillation can produce a series of activity patterns, instead of many repetitions of the same pattern. The continuous strengthening and decay of synapses due to recent use and disuse entails that any widespread stimulation of an area would tend to bias this competitive, sequential pattern generation to a replay of patterns that were recently active. Thus, if column $C - 1$ had recently produced patterns X, Y, Z, the signal pattern Q coming down from column C could initiate gamma waves that reactivated each of X, Y, and Z in quick succession, that is, with a period of 10–25 milliseconds.

As shown in figure 6.13, this fast replay brings the temporal lag between patterns down to within the window necessary for spike-time-dependent plasticity (STDP): 0–50 milliseconds (Fregnac 2003). Hence, even if the durations and temporal gaps in X-Y-Z are much longer (i.e., seconds or more), the gamma-driven replay compresses the series into a range where Hebbian learning can link the patterns together such that, in the future, they can stimulate one another. Thus, a single signal could initiate the entire sequence: it becomes self-completing within column $C - 1$ when given a small push. Column C, the receiver, would learn the triggering pattern of X, Y, and Z: an integration of their signals, all of which arrive on the same active (rising) phase of its slower oscillation (driven by layers 5/6 but also affecting layer 4). This integrated signal, Q (recorded as an activation pattern in layer 4 of column C, then sparsified to pattern Q_3 in layers 2/3, and finally integrated into pattern Q_6 in layers 5/6), serves as the abstraction of column $C - 1$ in column C. Due to the omnipresence of Q_6 during each of X, Y, and Z, Hebbian synaptic tuning in the dendritic mat of column $C - 1$ should ensure that the future presence of Q (Q_6) can trigger X-Y-Z.

As part of the same or a similar task, column $C - 1$ may also execute series X', Y', Z', which abstracts to state R (and R_3 and R_6) in column C; and yet another series may abstract to pattern S. Then, the sequence Q_3-R_3-S_3 in layers 2/3 of column C could, via beta-gamma oscillations, get abstracted into another pattern (K) in column C+1, and so on. Building such an abstraction hierarchy using only a few columns could prove difficult, but given that the primate brain contains millions of cortical columns, it is not hard to imagine large groups of columns organized into clusters, all of which (more or less) feed forward into another such cluster, and then on to another cluster for maybe a few hundred or thousand iterations. Then, if each cluster employed this basic beta-gamma callup procedure, a gradual chunking and abstraction process would seem possible.

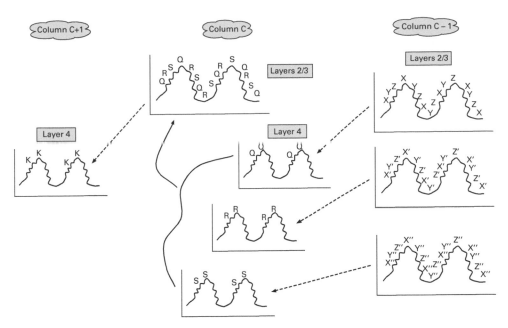

Figure 6.13
Using beta-gamma replay to chunk and abstract activation patterns. Sequences X-Y-Z, X'-Y'-Z' and X"-Y"-Z" play out in column C − 1 at different times. Beta stimulation from column C (layers 5/6) invokes beta-gamma waves in layers 2/3 of column C − 1, which replays a recent sequence, for example, X-Y-Z. In layer 4 of column C, the beta-gamma wave is processed at the slower beta frequency, thus integrating several signals before expressing a firing pattern, for example, Q, for a few gamma cycles at the top of each beta cycle. Other series in column C − 1 yield abstractions R and S in column C. Later, when layers 2/3 of column C receive beta stimulation from column C + 1, they produce a beta-gamma oscillation over patterns Q, R, and S, which, in turn, become abstracted to pattern K in layer 4 of column C + 1. To avoid notational clutter, this diagram denotes an activation pattern with a single letter, such as Q, regardless of the layer in which it appears, with the understanding that the same abstraction will invoke different patterns in each layer of a column.

This general mechanism mirrors concepts such as fixed action patterns (Llinas 2001) and hierarchical models of the brain (Ballard 2015), wherein higher brain regions send simple enabling signals to lower layers, which then perform intricate series of operations, without additional guidance. Sensory inputs predicted by the top-down signals are simply cancelled out, thus not tasking the system with unnecessary information transmission. Once again, this illustrates the inside-out view of cerebral control (Llinas 2001; Buzsaki 2019), since the high-level expectations run the show, only to be interrupted by unexpected inputs.

Buzsaki (2006, 2019) also champions beta-gamma callup as the means by which the hippocampus and neocortex exchange patterns. During exploratory activity, HC *calls up* information from the cortex by sending theta waves (4–7 Hz) to the cortex, which responds with (the higher-frequency) gamma waves (bundling many events into the active (rising) phase of one theta wave). The hippocampus then processes and (temporarily) stores these aggregates via STDP. Then, during non-REM sleep, the sleep phase with the greatest decoupling of brain and body, the process reverses and the cortex calls up HC with delta waves (0.5–3 Hz), and HC responds with ultra-fast (100–200 Hz) ripple waves, which pack several dozen patterns into the active cycle of one delta wave, thus enabling more permanent storage of the associations among these patterns in the cortex.

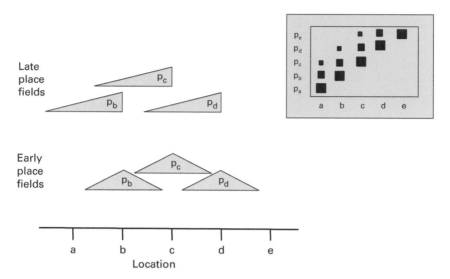

Figure 6.14
Phase precession in hippocampal place cells, $p_a - p_e$., with a–e on the lower axis denoting locations along a path. (Left) Triangles represent the place fields for their corresponding place cell, with height coarsely reflecting firing level (when the rodent is located at the spot on the horizontal axis). During early trials, the place fields are symmetric, with strong activation at the target location and gradual tapering of activity both before and afterward (in time / space), assuming the rodent moves left to right. After many trials, asymmetric fields form in which the cell fires several locations prior to the target but few (or no) locations afterward. (Right, inset) Predicted futures for well-trained place cells. Columns convey place-cell activity when rodent resides at location on the x axis, for example, when at location c, place cells p_c, p_d, and p_e are strongly, moderately, and slightly active, respectively, as reflected in the sizes of the dark rectangles.

Phase precession in the hippocampus

Prediction, oscillations, and the hippocampus come together in another, related, manner involving the phenomena of *phase precession*, as revealed by Mehta (2001). Many pyramidals in CA3 and CA1 have been identified as *place cells* (Andersen et al. 2007), meaning that their firing rates are strongly correlated with the locations of the animal: when at or near location X, the place cell(s) for X will show high activity; otherwise, they will fire infrequently. (See chapter 3 for more details on place cells.)

Mehta's neurophysiological studies show that as a rodent becomes familiar with a particular path, for example, $a \rightarrow b \rightarrow c \rightarrow d$, it will establish place cells p_a, p_b, and so on for each location. These neurons gradually become *predictive* in the sense that p_d will begin firing well before the animal reaches location d. In other words, the receptive fields (aka place fields) for the place cells extend backward in time and space, as shown in figure 6.14.

This has a relatively straightforward explanation in terms of spike-timing-dependent plasticity (STDP) and oscillations. As a rat progresses along the path, the corresponding place cells will fire, in sequence, with each place cell showing activity a bit prior to, during, and a bit after the visitation of its corresponding location, that is, the place field is symmetric about the location. The time delay between visiting locations a and b may be significantly greater than the normal window for STDP (40–50 msec), so connections between p_a and p_b cannot strengthen as a result of normal, sensory-induced activity. However, these place-cell firing sequences are often repeated after (and before) the rodent's search process at a much higher (gamma) frequency (40–100 Hz), thus bringing the sequential firing within the STDP window. Hence, p_a may fire a mere 10 msec before p_b, thus leading to a strengthening of the $p_a \rightarrow p_b$ synapse (and similarly the $p_b \rightarrow p_c$ and $p_c \rightarrow p_d$ synapses). In addition, this should also enhance $p_a \rightarrow p_c$ and $p_a \rightarrow p_d$ although to a progressively lesser degree due to the longer (albeit compressed) delays.

Conversely, the gamma replay, by invoking p_b activity immediately after p_a, will promote depression of any $p_b \rightarrow p_a$ synapses via the other half of the STDP curve, that is, when postsynaptic firing precedes presynaptic firing. Thus, the succeeding place cells lose the ability to stimulate their predecessors, while the predecessors have gained the ability to excite their successors.

The net result, after many trips along the path and many rounds of gamma replay, is a collection of *asymmetric* place fields in the sense that p_x becomes active for many of x's predecessor locations but for few or none of its successors. As a whole, the CA3–CA1 region becomes predictive, because when sensory input indicates location x, the place cells for several successors to x will fire. When the place *fields* skew backward in time/space, the place *cells* fire predictively, indicating the locations that the animal will soon visit, with the highest activity pinned to the most immediate successors.

In Mehta's work, the strengthened and weakened synapses are those from CA3 to CA1 such that both regions initially house symmetric place cells for the same location, x, but the place fields in CA1 become asymmetric and predictive via STDP. Hence, CA1 activity represents a prediction based on excitatory connections from CA3. This can only reconcile with our earlier description (in chapter 3) of CA1 as a comparator if the predictive firings in CA1 trigger inhibition such that the anticipation is effectively subtracted from sensory reality arriving from EC.

In this scenario, note that the overall timing relationships could adapt to essentially satisfy predictive coding. For example, the sensory reality encoded by EC at time t travels to DG, then through CA3, and then to CA1, which thereby encodes a prediction for some future time $t + \Delta t$. Assuming that this chain of processing takes τ msec, then the prediction will appear in CA1 around the same time that the direct line from EC conveys the updated reality signal, for time $t + \tau$. If the hippocampal loop can calibrate such that $\tau \approx \Delta t$, then the prediction and reality signals in CA1 should correspond to the same time point, thus making the comparison and resulting prediction error more appropriate.

Evolution seems to have invented the hippocampus quite early in the game, with it always playing a very fundamental role in memory formation and receiving the most abstract information that a particular brain had to offer. Only in mammals are those HC entry signals distant, abstract remnants of the original sensory data, but in amphibians, reptiles, birds, and others, the HC can still play the roles of (a) generator of high-level, temporally extensive, prediction error; (b) detector of novelty; and (c) initiator of widespread synaptic modification. In fact, hippocampal structures appear to have arisen independently (an example of convergent evolution) in birds and mammals (Striedter 2005), which indicates a genetic tendency to produce and a selective pressure to possess exactly this type of circuitry. The HC's well-documented role in navigation (McNaughton et al. 2006; Burgess and O'Keefe 2003;

Andersen et al. 2007; Buzsaki 2006) further underscores its importance atop the functional neural hierarchy.

6.3   A Brief Evolutionary History of the Predictive Brain

Regardless of whether the earliest organisms acted with or without sensing, motricity is a crucial factor in any discussions of prediction and its origins. As discussed earlier, a moving organism gains a survival advantage from prediction, even if only of the procedural variant: it behaves *as if* it can foresee its immediate future, by, for example, various gradient-driven activities. Even bacteria have this ability, so primitive multicellular organisms probably developed similar talents as their motoric sophistry increased.

Oscillations also enter the scene quite early, since rhythmic motion among a suite of myocytes enables a structuring of the local environment (in terms of features such as controllable / predictable flow patterns) and bodily coordination. However, this typically requires inhibition of one form or another: chemical, electrical, gap junctions, interneurons, and so on. As shown above, when it comes to the neural signaling level, the most common excitatory and inhibitory neurotransmitters have nearly identical chemical structure. Thus, inhibitors probably evolved quite easily once excitors had arisen, or vice versa.

Inhibitory interneurons and the inherent delays of chained neurons also support the calculation of simple temporal gradients; and these combine with the signal-integrating features of both individual soma and recurrent subnetworks to provide the rudiments of control and prediction. So as neural circuits added even a modicum of computational distance between sensors and motors, the possibilities for prediction and control would have emerged naturally. In addition, the advent of synaptic connections between neurons paved the way for learning via pre- and post-synaptic modification.

Inhibition also facilitates comparator neurons that receive excitatory signals from one source (e.g., sensory reality) and inhibition from another (e.g., a goal or prediction). The comparator's output then represents a deviation or error, which then promotes changes to expectations, actions, or even the goals themselves. Early on, in more primitive nervous systems, any activation patterns that somehow represented imagined states (of the brain-body-world coupling) probably served as goals, to be compared to reality, with the resulting error governing behaviors that reduced the goal-reality gap. The neural machinery probably existed to support functions similar to those of a conventional PID controller.

The next evolutionary step, to something akin to an adaptive controller (as discussed in chapter 3 in relation to the cerebellum), would require a comparison of predictions to relatively raw sensory signals, with that difference then serving as a perception. Then, a second comparison would involve the goal and perception, with the resulting error governing overt behavior. For example, when running to catch its prey, the predator's raw reality signal includes considerable vibration due to its own body movement, which the predator's brain can predict. That expectation constitutes noise, which, when removed from the raw signal, yields the proper percept (of a fast but otherwise steady prey trajectory). And that percept, not the noisy raw signal, is the more reliable representation to compare to the goal state (of predator meeting prey) in the course of choosing an action: in reaching out a limb to grab prey, the predator benefits from steady, non-jerking movements (with steadiness defined with respect to the ecocentric reference frame), despite the gyrations of the raw signals.

The basic adaptive control circuitry constitutes a module that could be duplicated and differentiated many times, quite possibly mirroring the proliferation of cortical columns in higher mammals or even the expansions of the cerebellum. This creates a hierarchy (or heterarchy) replete with predictions, reality signals, and comparators. At higher levels, the distinction between expectations and goals might blur: an activity pattern within a given module might constitute either, depending on the firing intensity of the neurons and the attention given to them, that is, the gains on their outgoing connections. Thus, goals would be patterns with higher activations and downstream influences, and thus be more likely to affect action; goals are the *envisioned states* that the brain will more stubbornly seek to achieve, while predictions have less causal force and may function as more *wishful thinking* (for positive expectations) and *traps to avoid* (for negative predictions).

Levels of this heterarchy that are somewhat distant from the peripheral sensory and motor apparatus would receive more delayed and multimodal signals, thus precluding them from low-level monitoring or control. Instead, their activation patterns would constitute abstract representations that encompass larger swathes of space and time while still playing a predictive role. The hippocampus seems to occupy the top, most abstract, tier of this organization.

Along each of these branches of the evolutionary tree, the constraints of energy conservation, firing-rate stabilization, and signaling efficiency would have incentivized proliferent inhibition and predictive coding. During development, the need to strike proper balances between competition and cooperation, and between hypo- and hyperactive firing rates, would provide selective advantages to networks with the right mixture of excitation and inhibition. Neural Darwinism (Edelman 1987) and its stipulated *survival of the best networkers* would encourage an individual neuron (N) to synchronize firing with local neurons (i.e., cooperate) to increase the chances of promoting downstream firing (more cooperation), and thus solidifying efferent synapses (and fortifying N's immediate network) via long-term potentiation. Conversely, it would also motivate N to inhibit some neural groups, to reduce competition for targets. Hence, an axon might invade some regions and synapse on excitors while breaching others to stimulate the local population of inhibitors. This *invasion with inhibitory intent*, a blatant act of competitive sabotage, would fortuitously form the key negative influence of a higher group's prediction on a lower group's comparator. Thus, predictive coding networks could emerge from the dueling forces of competition and cooperation driven by the basic needs of all neurons to survive and maintain relatively stable firing rates.

In parallel with the elevation of predictive topologies, oscillation gradually becomes encephalized. From a crucial factor in myocyte coordination, it ascends the heterarchy with each newly evolved level, providing a tool for synchronization of neural firing (and the ensuing cooperative and competitive benefits) while also promoting intraregional synaptic change and interregional information transfer, such as between the hippocampus and neocortex. This very simple mechanism has extremely far-reaching consequences in the emergence of intelligence.

Finally, facilitating each of these variations (i.e., each evolutionary branch) are the fundamental components of Kirschner and Gerhart's (2005) theory: weak linkage, modularity, duplication and differentiation, and the grow-to-fit dynamics of neural development. Each of these features, clearly evident in the brains of most organisms, provides a robustness and flexibility to life such that random genetic mutations have more neutral and positive consequences than any statistical random-walk analysis would predict. There does seem to be a certain inevitability to the emergence of predictive machinery when one buys into the theory of facilitated variation and realizes how neural systems and their development so perfectly instantiate each of its core tenets. From a billion-year evolutionary perspective, the capacity to generate expectations is certainly expected.

7 Evolving Artificial Predictive Networks

7.1 I'm a Doctor, Not a Connectionist

In a classic episode of *Saturday Night Live* (Shatner 2013), William Shatner—who played the starring role of Captain Kirk in the long-running television series *Star Trek*—is a guest speaker at a convention of ardent fans of the program, known (not so affectionately) as Trekkies. They ask him several detailed trivia questions about the show, questions about things that only a dedicated Trekkie would care to remember. Shatner draws a blank on all of them, much to the puzzlement of his audience. Finally, Captain Kirk lashes out at the crowd, uttering the famous line: *Get a life!*, followed by stern encouragement to move out of their parents' basements. He then storms off the stage and tussles with one of the conference hosts. However, he soon returns, somewhat mollified, and explains to his stunned followers that the whole thing was just an act, a re-creation of a *Star Trek* episode in which Captain Kirk temporarily becomes evil. He exits the podium amid admiring applause, and everything is as it should be in the Trekkie orbit.

I often imagine that groundbreaking scientists, those who produce actual paradigm shifts in their fields, must occasionally feel like William Shatner when they attend large conferences and see hundreds or thousands of extensions and spin-offs of their famous work being presented by the rest of us, the people doing what Thomas Kuhn calls "normal science" (Kuhn 1970). Most of this work probably warrants the label *derivative* from the harshest critics, but the renowned paradigm shifters and normal scientists alike recognize the importance of these further investigations into the applications, scope, merits, and drawbacks of the original breakthrough.

However, at some point, even the revered pioneer must recognize that *things have gone too far*. Their *intellectual hammer* has, in the eyes of their followers, converted all the world's problems into nails, creating a conceptual and methodological myopia that serves no significant purpose other than to produce more publishable documents in obscure journals.[1] This is when I can envision these pioneers taking the stage and shouting, *Get a few plausible alternatives!* to their fans. I then awaken from my daydream and get back to my normal science, because I need the publications and can never quite unleash that next paradigm shift.

Sir Geoffrey Hinton has revolutionized the field of artificial neural networks, and in turn, AI itself, although his recognition and rewards took decades to materialize. He was one of the inventors of the backpropagation algorithm (Rumelhart, Hinton, and Williams 1986) and of several other network architectures and algorithms, some of which were detailed in earlier chapters. Few breakthroughs in connectionism lack his immediate or indirect influence. As also discussed earlier, the core of backpropagation involves computing the gradients of a network's outputs with respect to most (if not all) of the networks modifiable parameters, which can easily number in the millions or billions in the deep-learning applications of the 2020s. Many, many problems seem to have become the nail to the hammer of the long-distance gradient, often for good reason, but sometimes as an awkward stretch.

The closest that Hinton has come to a Shatner-Trekkie moment may have been a 2017 interview (Levine 2017) when he admitted a *deep suspicion* of backpropagation and gave the following recommendation: "My view is throw it all away and start again.... I don't think it's how the brain works.... We clearly don't need all the labeled data."

In this and other interviews, he suggests that the next generation of researchers will need to devise new breakthroughs for pushing AI forward. However, as recently as 2020, Hinton and fellow researchers (Lillicrap et al. 2020) continued to propose theories as to how the brain may perform backpropagation. Captain Kirk has returned to the stage to give the Trekkies renewed hope, although this mixed messaging confounds researchers at the intersections of connectionism and computational neuroscience.

Despite all of the recent successes of deep learning, it has severe limitations in terms of the understanding (or lack thereof) that these systems have of the domains in which they learn (by processing many, often millions of, sample input-output pairs). That comprehension is often shallow, based primarily on syntactic similarities between the examples, whether pictures, text, or sensory data.

Stories abound (Mitchell 2019; Larson 2021) of deep nets trained to differentiate images of, for example, living organisms from inanimate objects. Although one might surmise that a successful neural-net classifier of such images had somehow achieved a deep understanding of life by viewing millions of pictures, what it really learns is that the background for the inanimate objects is often a table or wall with simple patterning, while the plants and animals, photographed in the wild, have intricate backdrops (or blurred backgrounds from a camera focused on the chipmunk on the limb of a tree with mountains miles off in the distance). Even more revealing are examples in which a few *meaningless* pixels are mutated, but the system changes its classification from, say, *canary* to *school bus*. Although more careful attention to visual context can ameliorate the former problem, nothing prevents the network from springboarding off of other superficial patterns to achieve excellent classification accuracy while still lacking the basic understanding of the world that most two-year-old humans possess.

In general, backpropagation seems to be a very attractive local maximum in the landscape of learning algorithms, and one that is hard to venture away from without plunging into valleys of mediocre performance. However, higher peaks probably exist, and it will take pioneering work, with no guarantees of (even mild) success, to find them. Recently, in the period from approximately 2010 to 2020, the safer research expeditions normally include backpropagation in their backpacks, thus guaranteeing at least reasonably competitive results (and a publication or two).

Although much less publicized than the deep-learning achievements, more ambitious attempts to leverage biological principles for AI gain have abounded over the past several decades. Many of these more biologically plausible approaches involve parallel search in the form of artificial evolution, and a subset of those include simulated development as well. As described below, these methods continue to hold promise, particularly when combined with predictive coding.

7.2 Evolving Artificial Neural Networks (EANNs)

For several decades after the popularization of backpropagation (in the mid 1980s), the numerous weaknesses of the approach motivated researchers in the evolutionary computation (EC) community to hunt for better, non-gradient-based methods for training neural networks. These often-cited (prior to around 2012) weaknesses of backpropagation included a tendency to get stuck at local optima, an innate hunger for exorbitant amounts of data in the form of input-output pairs, slowness due to the need to process each such pair thousands of times, vanishing gradients (particularly in recurrent networks), and a general lack of biological realism (which was of debatable relevance in discussions between engineers and neuroscientists).

Those researching evolvable artificial neural networks (EANNs) used these weaknesses, particularly that of premature convergence to local optima, to champion alternate approaches based on evolutionary algorithms. In these, genomes typically encode all of the networks weights or an algorithm for generating those weights; these embody direct and indirect genomic encodings, respectively. The parallel search manifest in evolutionary computation helps avoid premature convergence, thus producing weight vectors close to the global optima: their networks achieve optimal performance on the focal task, whether standard classification or autonomous goal achievement in a simulated world. Several interesting overviews of this subfield of EC have been published (Yao 1999; Stanley et al. 2019; Baldominos, Saez, and Isasi 2020; Downing 2015). Typically, these methods, whether direct or indirect, produce a fixed set of weights for the phenotype (network) from the genotype (a long string of bits, integers or reals). Unfortunately, this also lacks biological plausibility, since it is now well known that the genomes of even relatively simple animals cannot possibly encode the strengths of all synapses in the mature brain. Those efficacies must be tuned by a combination of genetic (nature) and environmental (nurture) factors during prenatal development and postnatal growth and learning.

Indirect encodings span a wide spectrum, from those that (simply) embody algorithms for generating weights to those that simulate a complex developmental process that governs the formation and interconnection of neurons and neural groups/layers. Stanley and Miikkulainen (2003) provide a nice classification of these systems, many of which are detailed in *Intelligence Emerging* (Downing 2015). Unfortunately, the intricate developmental approaches rarely produce networks capable of reaching the upper echelons of performance on benchmark problems (Hiesinger 2021; Baldominos, Saez, and Isasi 2020), although they often yield brains for simulated agents that exhibit reasonably lifelike behavior in artificial life environments. These (frequently disappointing) results on the biologically impressive end of the indirect-encoding spectrum serve as a warning to very carefully consider each neuroscientific nuance before mixing it into a connectionist scheme.

As reviewed by Soltoggio, Stanley, and Risi (2018), a different but related class of EANNs, the evolved plastic artificial neural networks (EPANNs) have been investigated (also over several decades) to more closely mirror the natural process by which intelligence emerges. In EPANN populations, phenotypes may employ Hebbian learning, backpropagation, or even evolved weight-updating algorithms as their source of lifetime learning (aka plasticity). Hence, the weights encoded directly or indirectly in the genome serve as only starting points for experience-based synaptic change.

Around 2012, deep learning (DL) began producing exceptional results on very hard problems. The repercussions were so powerful and widespread that several fields, such as image and natural-language processing, were essentially upended. Conferences that previously summarily rejected neural-network papers began accepting nothing but deep-learning approaches. Areas such as artificial life, EANNs, and EPANNs had less trouble absorbing the DL shock wave, since neural networks were already common tools in these pursuits. Still, changes of focus were inevitable once it became clear that news of backpropagation's demise had been greatly exaggerated.

By 2012, most of the abovementioned backprop deficiencies had been successfully combated. For example, deeper analysis showed that the risks of getting stuck at local optima were actually quite low (LeCun, Bengio, and Hinton 2015), and the burgeoning World Wide Web started supplying an abundance of raw data (particularly text and images) for training networks. Performance issues became rather moot with the widespread availability of graphic processing units (GPUs) and tensor processing units (TPUs), which backpropagation can exploit due to a multitude of matrix operations and an easily parallelizable aspect of the algorithm's training regime. Vanishing gradients (the fact that many of the long-distance derivatives have near-zero magnitudes, particularly in multilayered and/or recurrent nets) were reduced significantly by both (a) a plethora of (remarkably simple) alternatives to the logistic and hyperbolic-tangent activation functions (Goodfellow, Bengio, and Courville 2016), such as rectified linear units (Glorot, Bordes, and Bengio 2011); and (b) rejuvenated interest in long short-term memory (LSTM) units (Hochreiter and Schmidhuber 1997), which employ multiple gating functions to sustain certain signals (and hence the gradients in which they participate) during recurrent processing of serial data streams such as text and video.

Hence, of these original complaints about backpropagation, only the algorithm's biological plausibility remains suspect. None of the improvements mentioned above add any notable neuroscientific realism to deep learning—quite the contrary. However, just as LSTMs for sequence processing experienced a rebirth in the twenty-first century, so too did convolution networks (Fukushima and Miyake 1982; LeCun et al. 1990; LeCun, Bengio, and Hinton 2015) for image processing, and these have deep biological roots in the architecture of the mammalian visual system. But for every biologically plausible or brain-inspired supplement to deep learning, a host of useful but nonbiological *pure engineering* solutions arose. Textbooks on deep learning, such as Goodfellow, Bengio, and Courville's (2016) popular overview, appeal much more to statisticians than to neuroscientists. And so, in the early 2020s, a rather ominous (and growing) chasm still separates deep learning from computational neuroscience.

7.2.1   Reconciling EANNs with Deep Learning

As mentioned earlier, the adaptations of the EC and ALife communities to the sudden rise in deep-learning prowess were not paradigm-shattering: many of the existing tools could be modified to include DL components. In addition, the computational advances that fueled DL's rise could also be leveraged in simulated evolution.

The recent work led by one of EANN's pioneers, Risto Miikkulainen (Miikkulainen et al. 2017), provides the ideal prototype for a post-2012 combination of EC and DL. Their comprehensive CoDeepNEAT system integrates several aspects of two classic EANN systems with a variety of DL's primitive modules and network architectures.

CoDeepNEAT reuses cornerstone concepts from the SANE system (Moriarty and Miikkulainen 1997), one of the earliest successful EANNs. SANE coevolves a population of neurons with a population of blueprints; the latter constitute templates for combining individual neurons. Fitness testing involves selecting a blueprint and then instantiating its components with randomly selected members of the neuron population to produce a fully functioning network, which is then subjected to a performance test, the results of which contribute to the fitness of both the blueprint and the instantiating neurons. In CoDeepNEAT, the granularity rises from the neuron level to that of the neural group such that blueprints now specify complete, multilayered, DL networks; and blueprint components are groups of network layers, called *modules*, as shown in figure 7.1.

Another popular EANN system (developed in Miikkulainen's lab) named NEAT (Stanley and Miikkulainen 2002), contributes one of its core mechanisms, speciation, to CoDeep-NEAT. In the original NEAT, and in its long chain of descendants, the population of neurons is partitioned into *species*, each of which can evolve somewhat independently of the others: individuals in one species do not directly compete with those in another. This often gives newly generated individuals in one group a fighting chance of survival and further evolution.

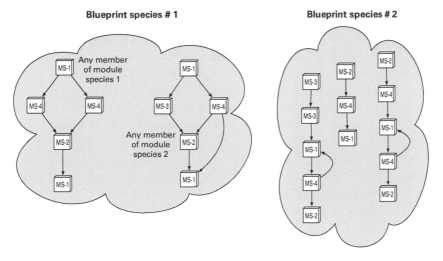

Figure 7.1
In CoDeepNEAT, blueprints provide templates for combining modules, the smallest atomic units, which are grouped into subpopulations denoted here as MS-i (the ith module species). Blueprints and modules coevolve in the system.

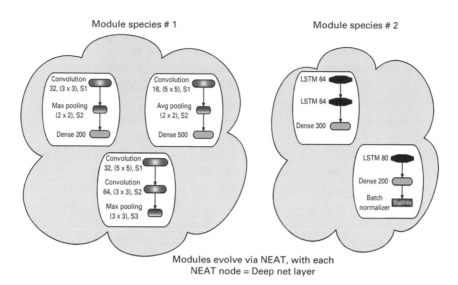

Figure 7.2
Species of modules in CoDeepNEAT, with each module consisting of one or more layers. The lowest-level primitives that evolution can modify in CoDeepNEAT are layers, not individual neurons.

Once again, CoDeepNEAT reuses a mechanism but lowers the resolution: whereas the atomic building block in NEAT is the neuron, in CoDeepNEAT, it is the neuron layer. Each module then consists of several layers, for example a convolution layer of one size feeding into one of a different size that outputs to a flat (aka dense) layer. Figure 7.2 shows two sample species whose individual modules incorporate many standard DL primitive layers: convolution, LSTM, max and average pooling, batch normalization, and so on.

Sewing all of this together clearly takes considerable human and computational resources, as indicated by the length of the author list for the original CoDeepNEAT paper (Miikkulainen et al. 2017) and the published run-times, respectively. As illustrated in figure 7.3, the resulting system performs fitness testing (far and away the most computationally demanding aspect of EANN operation) by instantiating blueprints with modules to form complete deep nets, which then go through training and testing phases to yield a performance score, which then contributes to the fitness of both the blueprint and the individual modules.

Although unable to beat the best hand-designed (often by large groups of dedicated researchers) deep networks on popular benchmark classification tasks, CoDeepNEAT performs very well on a wide range of data sets, including CiFAR-10 (Krizhevsky 2009), Penn Tree Bank (Marcus, Marcinkiewicz, and Santorini 1993), and Omniglot (Lake, Salakhutdinov, and Tenenbaum 2015). In general, it evolves very sophisticated architectures that would otherwise require hundreds or thousands of man-hours to design and test.

Automated parallel search, via evolution, may have trouble competing with backpropagation for finding the properties of individual neurons or synapses, but DL's diverse and complex primitives (i.e., layers with many parameters) have spawned an expansive and rugged design landscape (of multilayered network topologies) that requires the full power of automated search to navigate. Equipped with simulated evolution and big stacks of GPUs, Miikkulainen and others (listed in Baldominos, Saez, and Isasi 2020) have eagerly taken on the challenge.

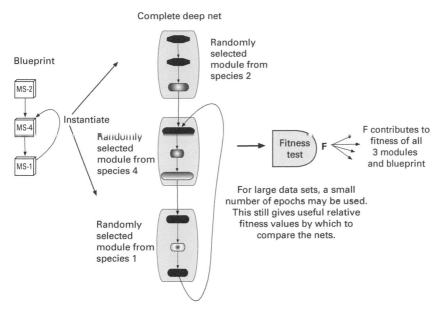

Figure 7.3
Fitness testing in CoDeepNEAT: Each individual network is the instantiation of a blueprint using randomly selected modules from the species that each blueprint component specifies. The performance score of the individual then contributes to the fitness of both the blueprint and the chosen modules.

Providing evolutionary search with even greater flexibility, Esteban Real and coworkers (2020) at Google Brain employ a variant of genetic programming (Koza 1992; Koza et al. 1999; Banzhaf et al. 1998) to evolve detailed aspects of machine-learning algorithms, with a slight bias toward those employed in gradient-descent approaches (though differentiation itself must evolve from lower-level primitives). Their system, AutoML-Zero, assumes only a fixed set of variables containing scalars, vectors, and matrices, along with a large collection of very primitive operations, such as dot product, max, min, mean, trigonometric functions, and so on. Evolution then finds equations linking these variables via the operations. Their ML algorithm enforces only the most basic structure: a setup phase followed by repeated iterations of a prediction and learning phase. The internal activities of each phase stem solely from evolutionary search. Thus, AutoML-Zero evolves topologies, forward-pass dynamics, and the parameter (e.g., weight) update computations.

This level of design freedom yields an enormous search space with only sparse regions of viable (let alone effective) algorithms. Real and colleagues employ Google's vast resources, running thousands of core processors over several days, to explore these needle-in-a-haystack design spaces. Their results are quite competitive on a wide variety of benchmark classification tasks, thus serving as another example of how hardware advances benefit many machine-learning approaches, not just conventional deep-backprop nets.

A host of diehard EANN researchers have not given up the fight to evolve individual synaptic weights, even as the sizes of weight vectors explode into the millions or billions. This general quest (and some promising results) show that the advantages of powerful distributed cores extend well beyond the DL community. To wit, evolutionary algorithms (EAs) are easily parallelizable—possibly more so than backpropagation. For most significant tasks, the overwhelmingly resource-hungry aspect of an EA is fitness testing of individual genomes, and the test of one genome normally occurs in isolation, completely independent of other genomes. Thus, each test can run on a separate core.

However, independence collapses at the population level, where reproduction occurs. At that point, the fitnesses of each genome are compared and/or combined to govern the choices of which parents will produce child genomes, how many, and, in some cases, how mutations and/or crossovers will occur. Once produced, the new genomes can then be sent to individual core processors for fitness testing.

The standard approach to EA parallelism has been to distribute fitness testing to the peripheral cores, but to then return fitness values to a central node that (a) contains all of the parent and child genomes and (b) can use the fitness information to compute the next generation of child genomes for export to the periphery. However, when genomes become very large, the communication costs become a significant factor. Hence, to exploit the full power of multiple cores, an EA needs an efficient means of transferring genomes. One approach to this problem (Salimans et al. 2017) (which is mirrored by several other projects), illustrates the continued competitiveness of evolutionary approaches against DL and deep reinforcement learning (DRL).

Salimans and his group at the Open AI Lab leveraged a parallel version of natural evolutionary strategies (NES) (Wierstra et al. 2014) to minimize the communication bandwidth between a central population-focused processor and the peripheral fitness-testing cores. As shown in figure 7.4, the NES algorithm generates children in a very *intelligent* manner, as opposed to the random mutation operators of conventional evolutionary algorithms. In a nutshell, NES generates several vectors of random mutations for each parent genome (P). It then applies each vector to P to produce a host of child vectors (C), each of which it then

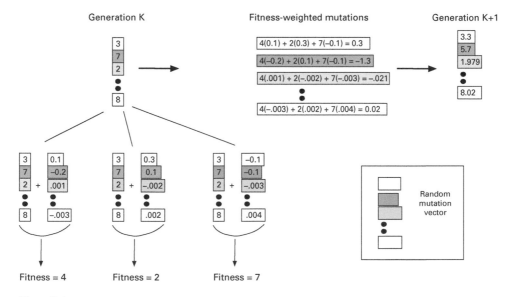

Figure 7.4
Basics of reproduction in the NES algorithm. This creates the child genome for generation K+1 via local search near the parent of generation K. Each branch of the tree represents one mutant of the parent, with all genes modified to some degree along all branches. Fitness values then weight the branch mutations to yield a single mutation that is applied to the parent, yielding a single child.

subjects to a fitness test. Now, instead of adding all of these children into the next generation, NES creates a single child by combining P with a new, intelligently crafted mutation vector consisting of the fitness-weighted average over C. Essentially, a local search is performed around each parent, but instead of taking the best path from P, the algorithm creates an average gradient as a guide. The next generation then consists of that single child genome.

To parallelize NES, Salinas' group essentially distributes the fitness tests of figure 7.4 to the peripheral nodes, along with a universal list of random seeds, as depicted in figure 7.5. That list enables each node to recreate the mutation vectors generated at all other nodes, a relatively simple process, while leaving the fitness tests associated with those vectors to the individual nodes. The scalar fitness values are then sent to a central processor, which gathers them and broadcasts the whole set to all nodes. Once each node knows all of the mutation vectors and their associated fitnesses, it can perform the same (simple) fitness-weighted averaging of figure 7.4 to produce the single individual of the next generation. Despite the redundancy of these calculations, they trade off well via a very minimal communication bandwidth: tens or hundreds of fitness values. A conventional, parallel EA for a large deep net might need a bandwidth of millions or billions of (binary, integer, or real) values for the transfer of complete genomes.

Salimans' group tested their parallel NES on a host of reinforcement learning (RL) problems, including a suite of fifty-one Atari games. They outperformed the classic work

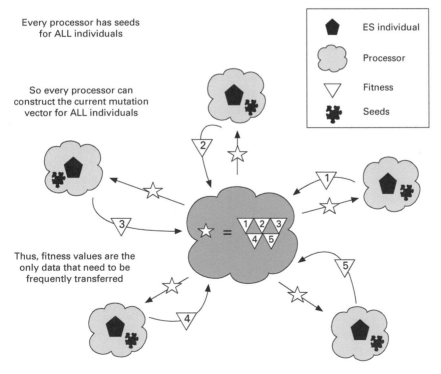

Figure 7.5
The parallel NES algorithm minimizes communication bandwidth by requiring each peripheral node to only receive fitness values, along with the single parent genome of the first generation and the list of random seeds used by each node. Since each node uses the same random-number generator, it can compute all random numbers generated by all nodes throughout the entire evolutionary search.

of Mnih et al. (2015)—who first showed that general-purpose RL using deep networks as function approximators could successfully approach and exceed human performance on a variety of tasks, thus illustrating a preliminary form of *artificial general intelligence*—on twenty-three of those games, while losing on twenty-eight. They achieved a speedup of two orders of magnitude when taking advantage of massive parallelism (1440 CPU cores spread over eighty machines) to successfully train a simulated 3D robot from the OpenAI Gym (Brockman et al. 2016): they reduced the training time from eleven hours (for a single machine) to ten minutes (for eighty machines). These competitive results against the top deep RL algorithms give one more indication that evolutionary approaches can exploit the same computational resources to keep apace of the impressive backpropagation-driven approaches.

Although sending a clear signal that EANN researchers have not given up the fight against DL's souped-up connectionist models, (and in many cases will seek a cooperative alliance by evolving that same brand of architecture), the work of Miikkulainen, Salimans, Real, and others only distance EANNs from biological plausibility.

The dual pathways to progress are clearly demarcated, though uncertain in their potential. Along one route, EANN work will continue to mirror DL in trying to incorporate each newly engineered DL concept or layer into EA genotypes and phenotypes, probably using the same argument about evolutionary search for tackling a complex architectural space. This route seems safe but clearly comes with huge computational costs, since simulating the training and testing of thousands of deep-net architectures can take days or weeks, even with a large stable of powerful cores. And, of course, just as the original backpropagation algorithm eventually ran into a wall that took several decades to overcome, the new, improved versions may eventually experience their own limitations. Already, the apparent lack of deep understanding in these networks constitutes a critical weakness (Mitchell 2019) that may hinder their safe deployment in mission-critical situations such as fully autonomous transportation, elderly care, geopolitical decision-making, and the like: areas that demand truly *general intelligence* of the human variety.

The other path sticks much closer to neurophysiology, evolutionary theory, developmental biology, and even psychology, in the belief that nature holds the pivotal secrets to success, the puzzle pieces of emergent intelligence. The remainder of this chapter explores this biologically plausible direction and its relationship to predictive coding.

7.3   Evolving Predictive Coding Networks

7.3.1   Preserving Backpropagation in a Local Form

Oddly enough, a split from deep learning does not require a strong denunciation of backpropagation: we need not throw out the backprop baby with the DL bathwater. A good deal of Hinton's mixed messaging on the topic stems from a continuing optimism that biological incarnations of the algorithm may exist, as explained by Lillicrap, Hinton, and their coauthors. They recognize but do not capitulate to the many criticisms of backprop (Lillicrap et al. 2020, 338):

But these arguments do not mean that backprop should be abandoned as a guide to understanding learning in the brain; its core ideas—that neural networks can learn by computing synapse-specific changes using information delivered in an intricate web of feedback connections—have now proved to be so powerful in so many different applications that we need to investigate less obvious implementations.

They characterize the main contributions in this direction as *neural gradient representation by activity differences* (NGRADs), which use neighbor-feedback connections that combine with local activities to yield an error signal, which then operates similarly to backprop's error term, despite being nonglobal and thus not as immediately representative of the network's overt performance. They point to Boltzmann and Helmholtz machines (as detailed earlier) as prime examples of NGRADs that exploit two-phase learning, such as contrastive Hebbian learning and the wake-sleep algorithm. This comparison across phases, and thus across time, requires neurons to cache states for later recall. As discussed in chapter 5, Bogacz and Whittington's version of predictive coding also achieves the many advantages of backpropagation, but without the long-distance gradients.

Both the NGRAD and predictive-coding alternatives involve very modular networks, with weak linkage between the levels / modules. Thus, they should provide flexible substrates for designing intelligent systems via AI search methods, such as simulated evolution.

7.3.2 Phylogenetic, Ontogenetic, and Epigenetic (POE)

As covered quite thoroughly in *Intelligence Emerging* (Downing 2015) and in several excellent review articles (Miller 2021; Stanley and Miikkulainen 2003; Stanley et al. 2019), EANN researchers have invested considerable effort in models of evolving and developing neural systems. These often invoke the POE acronym: phylogenetic, ontogenetic, and epigenetic, to cover evolution, development, and learning, respectively, or simply label them *EvoDevo* systems (which are somewhat agnostic about the epigenetic component). A notable relationship in these systems is between the P and the O. When a system uses artificial ontogeny in an attempt to actually *grow* phenotypes that achieve particular overt behaviors (e.g., solve control problems), a few different cases arise:

1. An optimal (or sufficient) phenotype (Z) is known ahead of time, so it serves as a fixed target for ontogeny. The researcher then desires a developmental process for *growing* Z, basically as a proof of concept that good (complex) solutions can be grown from simpler genotypic encodings. Hand-crafted solutions may be possible in these cases, although this is often a very difficult task even for a well-understood phenotype.

2. Z is unknown, and search will be required to find it. The researcher therefore needs a reliable tool to explore a (potentially immense) design space.

Case 2 may require only evolution / phylogeny, as in many classic problems (such as finding the weights for a predefined neural network topology) solved by evolutionary search with direct encodings, or it may require evolution and development (for design spaces replete with symmetries and other regularities that a growth routine could leverage), but it will rarely suffice to use development alone. Growth routines for unknown targets typically demand the support of a search procedure (and evolutionary algorithms often work best in these situations) to test many versions of the ontogenetic code. In short, when an AI researcher attempts to solve hard problems (with no known solutions), then they may enlist a P system or a PO system, but not an O system. Development almost always requires evolution.

The commonly cited advantages of artificial ontogeny (Downing 2015; Stanley and Miikkulainen 2003; Stanley 2007; Miller 2021) include

1. A complex phenotype can be encoded by a relatively small genotype, thus reducing the evolutionary search space. This assumes that the genome plus any predefined developmental routines suffice to produce a wide variety of useful phenotypes.

2. Phenotypes with important symmetries and repetitions are easier to evolve from simple genotypes combined with a recursive interpreter than from longer, more direct, genotypes that encode traits in a one-to-one fashion.

3. Long jumps in search space arise more readily from mutations to simple genotypes and recursive interpretation than from direct-encoding genotypes.

4. Indirect encodings are more biologically realistic. No organism's DNA can encode all of the details of their body or nervous system.

While the final point is obvious, it also seems unrealistic for even a recursive developmental routine to produce a full set of properly tuned synapses for a complex organism; lifetime learning appears essential for complex organisms, even though the simplest animals might get by with their *birth weights*. Hence, although PO systems (such as Stanley's popular HyperNEAT (2007) and Karl Sims' groundbreaking EvoDevo simulator (1994) produce complex phenotypic structures with fascinating behaviors, the route to sophisticated, flexible intelligence surely requires the E of epigenesis as well.

Soltoggio, Stanley, and Risi (2018) provide a comprehensive overview of the many PE and POE systems. What follows is a brief description of one such POE system, DEA-CANN (Downing 2007b), that I developed in 2007 with the primary goal of modelling key aspects of brain evolution, development, and learning while maintaining a relatively high level of abstraction. In particular, I sought a simplified model of neural development that captured the essence of the neuromeric model (Bergquist and Kallen 1953; Puelles and Rubenstein 1993; Striedter 2005), neural Darwinism (Edelman 1987, 1992) and displacement theory (Deacon 1998), but without the elaborate details of precursor cell duplication and differentiation, neural migration, and axonal growth.

The keys to this abstraction are evolvable binary tags / masks associated with each layer of neurons. The soma, axons, and dendrites of each neural layer use masks to influence connectivity and neuromodulator sensitivity. When the axonal tag of one layer complements the dendritic tag of another, the neurons of those layers have a greater chance of becoming pre-synaptic and post-synaptic neighbors; the somatic tag, when compared to the axonal tag of a pre-synaptic layer, determines the proximity of a synapse to the soma (and thus its initial, prelearning, strength).

These tag comparisons between layers yield an *invasion strength* for each layer relative to all others. As diagrammed in figure 7.7, layer 1's invasion strength for layer 2 is directly proportional to the axon-dendrite tag match between the layers and the duration of each layer's critical period (*devp time*), as well as the sizes of each layer, while the distance between layers 1 and 2 negatively affects invasion. All of these factors combine to manifest Deacon's displacement along with Finlay and Darlington's (1995) *late equals large* principle: the longer the critical period, the more chance for neural growth and axonal invasion.

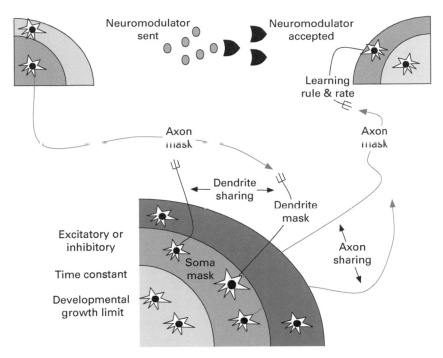

Figure 7.6
Overview of the main neurobiological details included in DEACANN, in abstract form. Each network consists of multiple neuromeres (semicircles) divided into concentric layers of homogeneous neurons. The evolvable neural properties of each layer include time constants governing developmental critical periods and firing dynamics, binary masks for dendrites, axons, and soma (which affect the connection probabilities between pairs of neurons in different layers), learning rates, neuromodulatory tendencies, and so on.

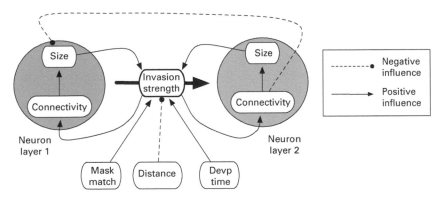

Figure 7.7
Main factors determining the invasion strength of one layer on another in the developmental process of the DEACANN system.

Neural Darwinism stems from the single negative influence from layer 2's connectivity to layer 1's size in figure 7.7. That link represents a set of relationships wherein the derivative of the size (S) of a region is positively correlated with the difference between its own connectivity (C) and the network-average connectivity ($\overline{C}$):

$$\triangle S_i = \alpha_g (C_i - \overline{C}) \tag{7.1}$$

where α_g is a small (e.g., 0.1) growth constant.

Connectivity, in turn, is the sum of a layers outgoing and incoming invasion strengths (I):

$$C_i = I_i^{in} + I_i^{out} \tag{7.2}$$

Hence, if the connectivity of layer 2 (C_2) increases, this can raise ($\overline{C}$) and potentially decrease $\triangle S_i$ for $i \neq 2$. This models situations where many layers have high penetration into layer 2, thus making it all the more challenging for any layer to invade.

In short, these equations capture key relationships between the sizes and connectivities of neural layers to reflect the basic tenet of neural Darwinism: the best networkers proliferate, while poorly connected neurons die.

Another central computational aspect of DEACANN is the three-stage process of translation, displacement, and instantiation. In translation, the (rather complex) bit-string genome is converted into a collection of neuromeres, each consisting of one or more layers, with each layer housing genetically encoded parameters, such as the tags, critical periods, and so on. The second, displacement, phase then constitutes an amalgamated abstraction of the theories of Deacon, Edelman, Finlay, and Darlington, with the sizes, invasion strengths, and connectivities of layers experiencing several iterative updates, but without producing any model neurons. The final, instantiation, phase then uses the final sizes and connectivities from displacement to determine the distribution of neurons within layers, along with the density and patterning of interlayer synapses. Additional genomic parameters determine the learning rule (Hebbian, anti-Hebbian, and so on) that applies to a layer's incoming synapses.

As with most POE systems, DEACANN produces interesting neural structures but little impressive behavior. When used to evolve controllers for a simulated starfish (see figure 7.8), it does create brains that learn to coordinate the five limbs to maximize forward movement, but artificial general intelligence is quite a long crawl away. Once again, the gap between computational neuroscience and AI is wide and deep. Although the inclusion of artificial ontogeny has interesting bridging potential, it remains largely theoretical and speculative as of this writing.

With very few exceptions, even the full PO and POE models fail to capture the essence of constructivism (discussed in chapter 6), since their developmental processes are insensitive to the task environment. In most EANN attempts at artificial ontogeny, evolution produces a genotype that either represents a full routine or pivotal parameters for a predefined routine, which then runs to produce a phenotype (i.e., a neural network). In either case, the genotype-to-phenotype conversion runs in a vacuum, with no external influences. In POE systems, this typically entails that all learning happens after development: the latter and the former run in lockstep.

As described in Nolfi and Floreano (2000), the main exceptions involve evolutionary robotics scenarios where each agent begins life with a skeleton neural network (for basic sensing and acting) and an evolved set of parameters for a fixed developmental routine.

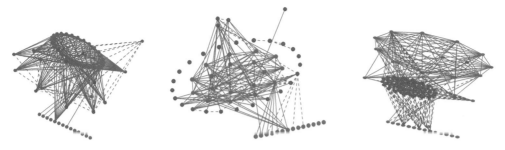

Figure 7.8
DEACANN-evolved neural networks for controlling a simulated five-limbed starfish.

As the agent wanders about and experiences the environment, its brain sprouts connections based on its evolved settings of the ontogenetic process and the activation levels of neurons, some of which reflect sensor or motor activity. In this way, the large-scale network architecture is molded by the sequence of postnatal interactions between the evolved ontogenetic properties and the environment. Although the tasks in these experiments are quite simple, they provide a nice proof of concept and several examples where identical genotypes exposed to different environments develop into divergent phenotypes.

Transferring these EvoDevo approaches to classic deep-learning tasks such as classification seems superfluous, despite the early success of the cascade-correlation network (Fahlman and Lebiere 1997) (discussed later in this chapter). DL has advanced light-years since then. Dynamically adding or removing neurons or connections to save a little RAM in the midst of supervised learning seems unnecessary in today's high-powered computing environments. The now-standard regularizing techniques (such as weight penalties, batch normalization, and dropout (Goodfellow, Bengio, and Courville 2016)) normally work well to prevent the overtraining and rote memorization that can result from an over-dimensioned network. In short, if the predefined, multilayered architectures and long-distance gradients of contemporary DL systems achieve success in a problem domain, then there is little reason to expect that P, PO, PE, or POE systems will outperform them. Since 2012, DL breakthroughs have made it quite clear that DL's version of E can get by on its own.

However, all approaches have their limits. Despite DL's current popularity, the lack of deep understanding in these networks—as revealed by dismal (and potentially dangerous) failures on adversarial examples—along with the burden of presenting each case hundreds or thousands of times during training, indicate that the road to AGI is not necessarily paved with convolution, dropout, rectified linear units, and long-distance gradients.

As a popular alternative, the *low road* to AGI features adaptive behavior and knowledge refinement via direct experience in a real or artificial world. This approach, popularized by AI luminaries such as Rodney Brooks (1999), Luc Steels (2003), Randall Beer (2003), and Jun Tani (2017), emphasizes situated and embodied agents (SEAs), that is, those that sense and act in an environment where they also have a physical presence (e.g., volume, mass, and so on) that needs to be taken into account during decision making. The underlying philosophy behind these approaches is that by *surviving* and problem solving in a rich environment, an agent acquires a commonsense understanding of that world, which it can then leverage to learn more advanced concepts via grounding in spatiotemporal experience, episodic-to-semantic induction, analogical association, and so on.

Originally, from the 1980s to early 2010s, SEAs were primarily designed using the tools of artificial life, such as neural networks, evolutionary algorithms, and finite-state machines (Langton 1989). In those cases, the neural networks rarely employed backpropagation but instead relied on evolution and/or Hebbian mechanisms to attain suitable weights and other parameters (such as time constants for individual neurons) (Beer 2003). The agents often adapted to perform well in their environments, but showed little evidence of deep understanding or commonsense knowledge.

In the 2010s, the field of reinforcement learning (RL) (Sutton and Barto 1998) skyrocketed in popularity on the coattails of deep learning. The combination of the two, known as deep reinforcement learning (DRL), led to many sophisticated game-playing and control systems, with AlphaGo (Silver et al. 2016) and its descendants (Silver et al. 2017; Silver et al. 2018) achieving what many viewed as a holy grail of game-playing AI: beating the world champion in the game of go. However, at their core, DRL systems rely on DL to recognize patterns in game boards, and although this produces a bevy of sophisticated state-action rules, it can still suffer from extreme brittleness. For example, a precursor (Mnih et al. 2015) to AlphaGo was highly touted as exhibiting AGI, because the same system could learn to play several dozen Atari games at or above that of top-notch human gamers. However, the brittleness of these systems clearly challenges any serious notions of deep understanding or AGI. For example, the self-trained system can masterfully play Atari's Breakout on a standard-sized grid, but if, post-training, the vertical position of the paddle changes by one pixel, performance deteriorates dramatically (Mitchell 2019). A legitimately, generally intelligent gamer could adjust to such changes instantly or over the course of a few games: she would require little or no retraining. Nature's brains do these things naturally.

7.4   Continuous Time Recurrent Neural Networks (CTRNNs)

The basic factor missing from the vast majority of ML, DL, and DRL systems is time. Vanilla backpropagation networks and many-layered convolution networks alike, even those with all the bells and whistles (dropout, batch normalization, various regularizers, and so on), ignore temporal aspects of the data: each paired input-target is handled independently of the next. Even DRL systems, which operate with temporal sequences of actions, tend to separate their data into state-value, state-action, or state-distribution pairs on which the supporting DL network then trains independently. In fact, they get better results by randomly shuffling these cases such that the states in a training batch of pairs are not consecutive (i.e., *correlated*) states in a problem-solving episode.

Although recurrent neural networks (RNNs) incorporate history, via feedback connections, they tend to handle time discretely, with each neuron operating on the same timescale. Real neurons, however, constitute a continuous dynamical system (i.e., brain) whose behavior spans many temporal scales, from milliseconds to hours and days. To capture this continuous, multiscaled property in simulation, computational neuroscientists often employ explicit time constants in their models. Some AI and ALife researchers take inspiration from these approaches, and although they often scale down the biophysical and biochemical details, they retain the detailed treatment of temporality in order to achieve more realistic behavior and better performance on dynamic tasks such as robot control.

Continuous time recurrent neural networks (CTRNNs) have become popular models for just such situations. Although used for many decades by several pioneers in neural networks (Grossberg 1969; Hopfield 1984), they gained prominence in more elaborate settings (i.e., as the brains for real and simulated robots) by the work of Randall Beer (Beer and Gallagher 1982; Beer 2003), Jun Tani (Yamashita and Tani 1996; Tani 2017) and colleagues.

The basic equation for a CTRNN neuron is

$$\frac{\partial u_i}{\partial t} = \frac{1}{\tau}[-u_i + \sum_k x_k w_{k,i}] \tag{7.3}$$

Here, u_i represents neuron i's membrane potential, while x_i denotes unit i's firing activity, which is often scaled to the range [0,1] via a logistic sigmoidal activation function (f). Thus, u_i maps to x_i as follows:

$$x_i = f(u_i) = \frac{1}{1 + e^{\theta_i - u_i}} \tag{7.4}$$

where θ_i represents a firing threshold for which higher values will shift the logistic function's S curve to the right along the x axis.

The strength of the connection from neuron k to neuron i is denoted by $w_{k,i}$, and the summation represents that of all weighted inputs. The $-u_i$ on the right-hand side of equation 7.3 indicates a *leak* term, wherein the fraction $\frac{1}{\tau}$ of u_i's previous value is lost, while the rest is retained—thus embodying a simple form of memory. That same fraction, of the sum of weighted inputs, can then be viewed as replacing the *forgotten* potential, $\frac{u_i}{\tau}$.

Temporality enters via the time constant τ, which controls the rate at which this replacement occurs. Lower values dictate drastic replacement, high frequency change, and thus a faster dynamics, whereas high τ signals a slower timescale wherein older values linger and new inputs have reduced influence on the neuron's current state. Note that when $\tau = 1$, equation 7.3 represents a neuron without local memory, and, essentially, a unit in a classic feedforward backpropagation network, whose current state is determined entirely by its current sum of weighted inputs.

A modeler can introduce τ into just about any neural network. It requires little change to the computations governing activity propagation but can have large consequences for learning (i.e., weight modification). For $\tau > 1$, memory is introduced, and this is equivalent to adding a recurrent connection from a neuron to itself. For backpropagation networks, this adds a whole new level of complexity to the learning phase, since it requires backpropagation through time (BPTT), whose details (Werbos 1990; Goodfellow, Bengio, and Courville 2016) are outside the scope of this book. However, as long as the network has a strict hierarchical organization, with well-demarcated inputs, outputs, and layer sequence, BPTT is a straightforward (though mathematically dense) process. BPTT handles layered networks with intraneuron and intralayer recurrence but becomes unmanageable in unstructured nets with no clear sequential structure. In many such situations, you may (eventually) get BPTT to run, but the learning will probably be disappointing.

7.4.1 Evolving Minimally Cognitive Agents

Randall Beer and his colleagues have a long-standing interest in small networks with full interconnectivity between internal neurons (such as in the Hopfield networks presented in chapter 4), but where each neuron could have its own values for τ and θ, and those values are not hardwired by the designer: the system must search for the proper values, which are very difficult to attain with BPTT. These networks realize abstractions of real brains by incorporating multiple timescales and thresholds, that is, a mixture of slow, medium, and fast neurons with logistic curves having different rise points spread along the x axis. These networks can thus respond to different temporal aspects of their sensory space and with actions of varying temporal scope.

Adapting the weights of these multiple-timescale networks is difficult for backpropagation but straightforward for an evolutionary algorithm. Beer simply evolved the τ_k and θ_k for each unit, k, along with all of k's incoming weights, $\forall i : w_{i,k}$ (the collection of which, for all k, constitute a genome) and then loaded them into a fixed network architecture, which became the brain of a primitive *bot / animat*, whose performance on a set task then formed the basis for the genome's fitness rating. The tasks were simple but carefully crafted to require *minimally cognitive* faculties, such as prediction of future states, memory for earlier inputs or previous actions, selective attention, and so on. The resulting animats were impressive, especially given the small size of the neural networks, often with no more than a half-dozen internal neurons, thus bolstering the status of the bottom-up ALife approaches to AI during a period (roughly 1990–2010) when other AI techniques struggled in dynamic environments.

As an example of evolving procedural predictive capabilities, figure 7.9 illustrates a classic object-tracking bot devised by Beer's group (Slocom, Downey, and Beer 2000). In one of the most challenging scenarios, the agent's whiskers act as proximity sensors, but they function only up to the point when the agent begins to move. The agent must then operate *blindly* for the remainder of the episode, using only the information that it has gathered prior to the sensory blackout. Thus, the agent needs to *predict* the spot (along the x axis) at which the object will hit the ground level ($y = 0$) and move itself to that location in time to meet / catch it. This agent has only six internal neurons, so actually forming an internal representation

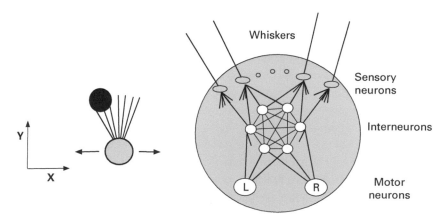

Figure 7.9
Simple object-tracking agent used by Beer and colleagues (2000) to display minimally cognitive predictive behavior. (Left) Agent (light oval) detects diagonally falling objects (dark oval) using whisker-like proximity sensors (thin lines) and can move left and right along the horizontal axis in trying to meet the falling object when it hits the ground. (Right) Topology of the agent's CTRNN: one sensory neuron per whisker, full forward connections from sensory to interneurons, which are fully intraconnected and have forward links to both motor neurons for left (L) and right (R) translation.

of the landing spot may be difficult—or, possible to form but difficult to employ as a *map* for guiding movement—so this is probably most aptly characterized as procedural prediction: the agent evolves to act *as if* it knows the location, since the well-adapted bot moves there.

In this and several other scenarios using similar bots and brains (Slocom, Downey, and Beer 2000; Beer 2003; Beer and Gallagher 1982), evolving a population of a hundred genomes over a few hundred or thousand generations (depending on the difficulty of the task) produced very successful networks. Although elucidating the secrets to success at the neural level requires considerable effort, Beer and coworkers have gone this extra mile in several publications (Beer and Williams 2015; Izquierdo and Beer 2013; Phattanasri, Chiel, and Beer 2007; Beer 2003; Beer and Gallagher 1982). A similarly detailed analysis of the blind object tracker has not been published, but we can speculate as to some of the basic mechanisms encoded in the six interneurons that would support procedural prediction.

Figure 7.10 shows how gradients and summations can once again enter the predictive picture, with delays enabling derivative computations, and with different subnetworks used

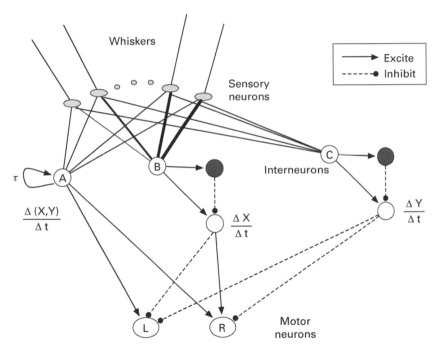

Figure 7.10
Hypothetical network for a whiskered agent that tracks diagonally falling objects. Each of the three interneurons, A, B, and C, detects a different feature of the object; dark ovals indicate delayed inhibitors. A well-chosen (i.e., large) time-constant (τ) allows neuron A to integrate sensor signals over several timesteps, thus giving a general indication of velocity: the more whiskers the object hits in a limited time, the higher its velocity. Neuron B has a gradient of incoming weights from the sensory neurons, as indicated by variable thickness of those connections. The triangle of neurons near B computes B's change over time, via delayed inhibition, which represents horizontal velocity, with a positive derivative indicating rightward motion. Thus, the derivative excites neuron R but inhibits neuron L. Neuron C requires no gradient of input weights; C's activation change over time (via another delayed inhibitor) estimates vertical velocity, which has a negative effect on both motor neurons: the faster an object is falling, the closer it will land to the current horizontal location. With appropriate synaptic weights and time constants, a CTRNN would probably need only two of the three detectors (i.e., less than seven interneurons) to solve the task.

to compute horizontal, vertical, and resultant velocity. As long as a few of these or similar mechanisms are realized by the six interneurons, a proper setting for the other weights would allow the agent to convert a time series of whisker signals into an effective action sequence. Although the interneurons are fully interconnected, evolving near-zero weights for many of the links would, in essence, *carve out* simple modules such as those in figure 7.10.

Since each whisker registers proximity, not merely contact, relatively rich information flows into the interneurons. Starting with neuron A in figure 7.10, it can provide a proxy for the resultant velocity by simply recording how many whiskers are hit during a short time frame, where the evolved time constant τ figures prominently in the calculation: it should be large enough to ensure the summation of whisker collisions over several timesteps. The output of neuron A would, in general, have a positive effect on both motor neurons, since a fast-moving object will require an equally speedy agent to catch it. However, the separation of vertical and horizontal speed also seems essential. An object with a dominating vertical velocity often requires little or no horizontal motion to catch, while the opposite situation of fast horizontal translation will demand fast movement, and possibly for a long distance. Thus, in the speculative network of figure 7.10, the vertical change component $\frac{\Delta Y}{\Delta t}$ has an inhibitory effect on both motor neurons.

The horizontal component $\frac{\Delta X}{\Delta t}$ is more complex. First, to recognize horizontal motion, the interneurons need some (at least partial) sense of spatial order among the whiskers. An object falling straight down to the agent along one (or a few) whiskers will promote an increasing sensory signal, which will produce high $\frac{\Delta Y}{\Delta t}$ activity, but that should not yield a high-magnitude $\frac{\Delta X}{\Delta t}$ output unless *different* whiskers activate on different steps. One primitive strategy for signaling these differences (with a very limited number of neurons) involves a graded set of weights, as depicted with the varying line thicknesses between the sensory units and neuron B. Now, an object moving left-to-right would produce a low input to neuron B, but each timestep should yield higher and higher inputs (i.e., $\frac{\Delta X}{\Delta t} > 0$). This would then excite the motor neuron for moving right but inhibit that for going left, as shown in the figure. Conversely, right-to-left motion would yield $\frac{\Delta X}{\Delta t} < 0$, whose value would then invert on its way to motor-neuron L, thus producing leftward motion.

With few exceptions, the successful evolved neural networks for minimally cognitive tasks do not perfectly fit the mold of diagrams such as figure 7.10: rather, they exhibit complex dynamics whose behavior can only be summarized by a very contorted logic. This is not the stuff of robotic engineering textbooks. But buried within the meshwork of weights, time constants, and biases, one often finds combinations of two or three neurons that perform recognizable functions. Identifying these small neural motifs and then tying them together into holistic functional explanations present some of the biggest challenges to the bottom-up approach to intelligence, even when the network contains a mere handful of units.

7.4.2   Cognitive Robots Using Predictive Coding

Jun Tani (whose seminal projects are summarized in his popular book, Tani 2017) sought to extend CTRNNs to humanoid robots, but the sheer size of the (visual) input and motor output spaces required more constraint on the neural networks: they could not wildly mix time constants and thresholds, and there were too many weights to find via evolution. Still, Tani recognized the importance of multiple timescales and wanted to mimic the spatiotemporal hierarchy found in animal brains, with higher-level processes typically dealing with greater abstractions (over both space and time). By enforcing strict layering on his networks, and assigning unique time constants to each layer, but allowing full intralayer recurrence, he facilitated adaptivity via BPTT: the proper weights could be learned in a supervised fashion.

The next trick is to find targets for supervised learning, and that's where prediction comes back into the story. As discussed in the introductory chapter, one cheap way to produce targets is to make predictions about future sensory or proprioceptive states, wait a timestep or two, and then compare them to reality (the target). Tani's networks predict both visual and proprioceptive futures and then use their deviations from reality, that is, the prediction errors, as the basis for BPTT. His networks learn weights, biases, *and* the initial state of each layer for each data sequence. Furthermore, they learn a special vector of latent parameters, known as the *parametric bias (PB) vector*, which serves as a bifurcation (i.e., switching) parameter for generating sequences. Tani often interprets the initial state of the highest layer as the *intentional* state for generating intended sequences, while the PB vector represents both an action and the expected sequence of sensory inputs that the action will incur. In this way, the PB vector governs both the top-down generation of actions and the bottom-up recognition of actions via the sensory portfolios they produce.

As depicted in figure 7.11, many of Tani's projects utilize a multi-timescale recurrent neural network (MTRNN) (Yamashita and Tani 1996), a multilevel version of the CTRNN with layer-specific time constants (τ): smaller time constants (faster dynamics) apply to the lower levels, while the higher levels run slower (higher τ). From this neuro-structural starting point, a functional hierarchy gradually emerges as the robot trains on particular tasks and learns via backpropagation. These networks exhibit self-organized chunking at the higher levels, wherein activity patterns in the upper echelons come to represent sequences of patterns in the lower, fast-dynamics areas. Thus, the abstraction can serve to both trigger and maintain the low-level series of actions and/or predictions. This emergent chunking of significant, task-specific activities helps validate claims by ALife researchers that abstract representations need not ground out in symbolic representations: a vector of real-valued sensorimotor activation levels may suffice.

However, as a testament to the versatility of Tani's approach, he has also used the adaptive PB vectors as a bridge between abstract patterns (i.e., *thoughts / intentions*) in one neural network and symbol-inducing patterns in a parallel, language-learning net (Tani 2014). Essentially, the intentions in the behavior and language networks gradually adapt to match one another, thus forming a high-level bind between the thoughts involved in doing a sequence of actions and those involved in communicating about it. This work provided one of the first truly impressive, biologically realistic, operational models of both the integration of symbolic and subsymbolic processing, and (perhaps most fascinating) the emergence of this link through real-world experience and adaptivity. The fundamental importance to artificial general intelligence (AGI) of these mental connections between action and communication clearly accentuates the significance of Tani's work for the future of AI.

Tani's contributions to prediction extend well beyond his method of gathering targets for supervised learning. Several of his projects (Ahmadi and Tani 2019; Matsumoto and Tani 2020; Queisser et al. 2021) deeply embrace predictive coding via the explicit use of Friston's free-energy principle (FEP) (Friston, Kilner, and Harrison 2006; Friston 2010) (explained in chapter 4). The basic premise in several of these projects is that neural systems adapt

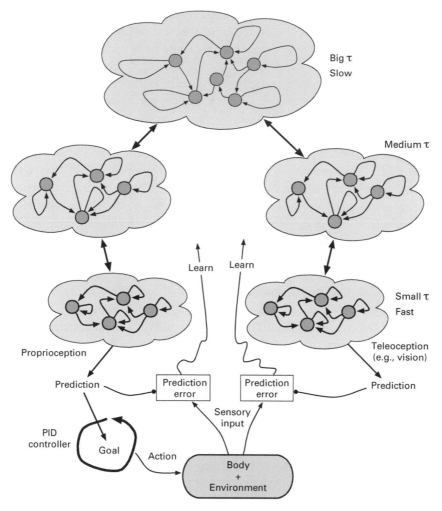

Figure 7.11
Basic neural network architecture and relationship to body and environment employed in several of Jun Tani's robots, based on similar diagrams in (Yamashita and Tani 1996; Tani 2017; Ahmadi and Tani 2019). Shorter, thicker interneuron connections denote stronger, more immediate influence, while shorter, thicker intraneuron loops represent faster dynamics. Prediction errors drive learning via BPTT. Note that actions are initiated as proprioceptive predictions that serve as goals for a PID controller, as seen in chapter 6's example of a spinomuscular regulatory circuit.

to minimize free energy, one expression of which is repeated (from earlier) with one small modification (the addition of ω) in equation 7.5.

$$F_g^r(d) = \omega \underbrace{D_{KL}(p_r(s|\Theta), p_g(s))}_{complexity} - \underbrace{\sum_s p_r(s|\Theta)ln[p_g(\mathbf{d}(\alpha)|s)]}_{(predictive)accuracy} \qquad (7.5)$$

For instance, Ahmadi and Tani (2019) instantiate the system-state variable s in equation 7.5 with the PB vectors of each level, which reflect the MTRNN's state. They then use $F_g^r(d)$ as an error function and compute gradients of it with respect to both network weights and PB vectors, using the (rather intricate) calculus behind BPTT (since the networks are recurrent). Learning then entails updating the weights and PB values to reduce free energy. [2]

They then do different runs of their robots using different values of ω, thus modifying the relative importance of accuracy and complexity, where the former denotes the net's ability to properly predict future sensory and proprioceptive states, while the latter quantifies the compatibility between the prior and posterior distributions of the PB vectors. With high ω, a robot prioritizes stubbornness: not changing its internal states despite confounding evidence. These states then cause the robot to follow through on its beliefs / intentions during its next round of prediction-making, by performing actions that have a high likelihood of producing its predictions. Conversely, a low ω encourages the agent to modify its beliefs to more accurately reflect sensory reality.

7.4.3 Toward More Emergent CTRNNs

Tani's work exhibits considerable emergence in the sense that useful functional building blocks and conceptual bridges arise (via learning) as weight and PB vectors. Given the fundamental constraint of the network's basic hierarchical topology with graded time constants, the most useful primitives emerge from BPTT; they are not given by the system designer. However, the success of the MTRNN and its descendants makes that general architecture an ideal goal / target network for more profound, biologically realistic examples of emergence in the POE framework.

First, let us designate Tani's family of neural nets as PN-BP: predictive networks that adapt via backpropagation. These architectures possess considerable biological realism, since they are abstract models of mammalian brains, with the upper layer playing the role of the prefrontal cortex, and lower levels mirroring sensory and motor cortices. The CTRNN and MTRNN also provide ideal support for the types of predictive networks / controllers discussed throughout this book, since they enable the free-form combination of excitation, inhibition, integration, temporal resolutions, and delays.

However, CTRNNs typically do not include development or learning, while the MTRNN does learn, but via backpropagation, thus forfeiting some biological realism. Figures 7.12 and 7.13 display several examples of extending the CTRNN / MTRNN framework with various combinations of P, O, and E so as to enhance the biological realism, and hopefully leverage some of nature's tricks.

One potential option is to employ NGRADs (see above) to achieve local (Hebbian and anti-Hebbian), biologically plausible, weight updates. These methods apply equally well to vanilla feedforward and recurrent networks (Whittington and Bogacz 2019): since all weight updates rely solely on adjacent pre- and postsynaptic activities and thus do not need to *wait* for backpropagated error terms, they can be performed any number of times during the recognition or generation of a data sequence. This constitutes a predictive network with biologically plausible learning (aka epigenesis), thus earning the acronym PN-E.

Another strategy is to fix the general architecture to Tani's MTRNN or to something that (in terms of structure) more thoroughly embodies predictive coding: a network such as the Helmholtz machine with predictions, targets, and comparators at every level. Adaptation would then involve evolution for finding the proper weights on all intra- and interlayer connections. Although this may not have been a very successful strategy when MTRNNs (let alone Helmholtz machines) were introduced, the advances by Salimans and colleagues

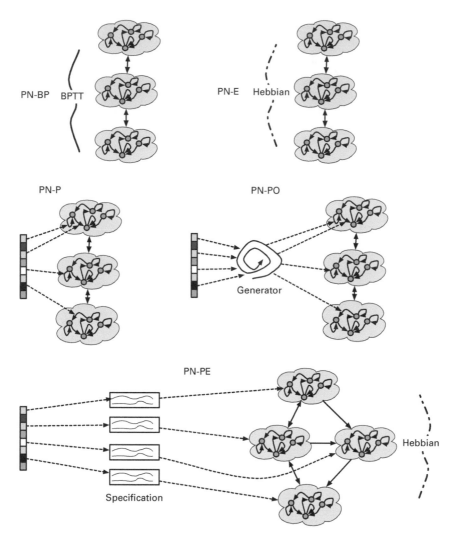

Figure 7.12
Various adaptive approaches to designing predictive networks (PNs) with a general architecture similar to Tani's MTRNN. P = phylogenetic, O = ontogenetic, E = epigenetic (Upper left) Basic approach used by Tani and colleagues: hand-designed topology with BPTT learning. (Upper right) Hand-designed topology with local, Hebbian-type learning. (Middle left) Genome (vertical, multitoned bar) encodes (fixed) weights for hand-designed topology. (Middle right) Genome encodes parameters for a generator that then runs to produce all (fixed) weights for a hand-designed topology. (Bottom) Genome encodes specifications such as number of layers, number of neurons per layer, intra- and interlayer connectivity patterns, and so on; weights adapt via Hebbian mechanisms.

(2017) (along with several other research groups) have made evolutionary search for (very large) weight vectors a competitive alternative. This class of solutions involves simulated evolution (phylogeny) as its sole adaptive mechanism, so we can abbreviate it as PN-P, despite the fact that evolving all synaptic strengths is no more biologically realistic than backpropagation. In the absence of development and learning, the PN-P system requires a direct encoding of each weight in the genome.

Although it's often just an issue of semantics, one can argue that any indirect encoding (i.e., one that lacks a 1-1 mapping from genes to network parameters), such as Stanley's

HyperNEAT (2007), is actually evolving a developmental (ontogenetic) recipe and would thus constitute a PO system. To produce a predictive network, a PN-PO system uses the genetically encoded developmental routine (or key parameters for a hard-coded routine), which then runs to produce connection weights. Potentially, the developmental recipe could also produce network layers, their units, and the connectivity patterns among them. Purists such as Julian Miller (Miller 2021; Harding, Miller, and Banzhaf 2009) would often argue that a developmental algorithm should begin from a single seed structure that duplicates and differentiates into the full phenotype (see below), but others prefer a more general interpretation of the concept to include any situation where a (possibly short) genotype runs through a generative process to produce a (possibly much larger) phenotype. Regardless of these philosophical differences concerning development, the key distinction between PN-PO and PN-POE systems is the former's lack of synaptic adaptivity: its generator produces fixed weights.

The next class to consider is PN-PE: predictive networks that adapt via evolution and learning. Again, these include MTRNN and purer predictive-coding architectural *schemes*, but various macro- and microlevels of structure now come under evolutionary control. For example, the number of MTRNN or Helmholtz layers, the sizes of the neural populations, each MTRNN layer's time constant, and so on. would be encoded in the genome. These have the same flavor as neural architecture search (NAS) techniques such as Miikkulainen et al.'s CoDeepNEAT (2017) (described above), but with the primitives specially tailored for prediction: comparators, delays, prediction-error transfers between layers, and so on. In addition, these systems would leverage synaptic tuning for *lifetime* learning, using biologically realistic mechanisms. If, instead, they employ conventional deep-learning techniques (as in most NAS systems), the more appropriate acronym would be PN-P-BP.

None of the above classes facilitate a creative design process that could craft something as complex as a hippocampus, cerebellum, or stack of cortical columns—or even very abstract versions thereof. Of course, other intricate neural topologies might work just as well, or better, but the search space may be so vast that cooperation among all three primary forms of adaptation might be necessary for designing brains for environments that demand sophisticated sensorimotoric and/or cognitive skills. In short, full PN-POE systems may be unavoidable in the long run, in the quest for artificial general intelligence.

Figure 7.13 portrays two of many possible PN-POE approaches. In these, the genome encodes several aspects of the developmental routine and a seed network configuration (denoted as *specifications* in the figure). For example, it might determine the number of layers and the initial number of units in each; development would then generate connections between units, initial weights, and/or additional units. In more extreme (and biologically realistic) cases, the genome primarily encodes a generator routine, which runs inside of one initial neuron or one simple layer. The ontogenetic recipe then directs duplication and differentiation of units, separation into distinct layers, and essentially all of the structural changes necessary for network formation. From the biological perspective, these extreme PN-POE approaches are the most interesting; but they present a formidable challenge to evolve adaptive networks that perform useful functions in nontrivial environments.

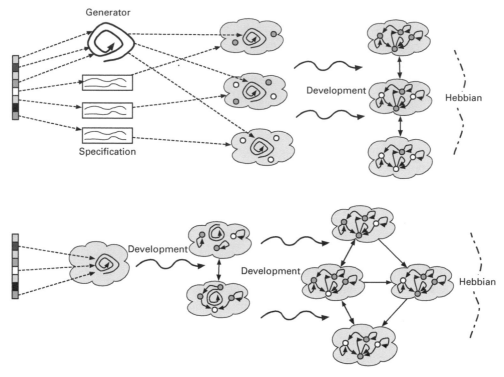

Figure 7.13
Two examples of adaptive predictive networks (PNs) that employ phylogeny, ontogeny, and epigenesis (POE). (Top)
The genome encodes both parameters for a generative routine and an initial set of conditions: the specifications of
the starting states for three layers, including the number and type of neurons. Running the generator then further
elaborates each layer, including intra- and interlayer connections. Hebbian learning handles lifetime adaptation.
(Bottom) A very biologically realistic PN-POE algorithm, where the genome encodes a generator, which then runs
to produce layers, neurons, and all connections and their initial weights, which Hebbian learning then tunes during
the agent's lifetime.

7.5   Predictive POE Networks

7.5.1   Simulating Neural Selectionism and Constructivism

As discussed earlier, narrow definitions of neural selectionism (Edelman 1987) and constructivism (Quartz and Sejnowski 1997) yield two disparate models of neural development.
In the former, overproduction of synapses prenatally (i.e., before birth) and neonatally (i.e.,
early in life) gives way to widescale pruning in a *competition to cooperate* among neuron groups, including sensory receptors and thus involving the environment to at least some
degree. In contrast, constructivism accentuates the influence of external factors on the gene-
sis of neurons and growth of *processes* (i.e., axons, dendrites, and synapses) throughout life.
A more flexible interpretation of either theory (and particularly of constructivism) admits
neurogenesis, process growth, and selective elimination of processes throughout a lifelong
development process that gradually decelerates with age. Thus, development and learning
do not occur in lockstep, but in tandem, with the former dominating early on and the latter
later, but with neither totally sidelined during a lifetime of mental adaptivity.

Modern deep learning (DL) provides a relatively close parallel to strict selectionism. The
network typically begins with full connectivity between layers (i.e., every neuron in the

upstream layer links to every node in its downstream neighbor), with all weights initialized to random values. Then, through backpropagation learning, the weights gradually diverge: some attain medium or large magnitudes, while others tend toward zero. Those vanishing weights have effectively been pruned by backprop, but practically (i.e., computationally) speaking, the weights still burden the system, since it continues to update their gradients through all training epochs. A more efficient approach might actually remove them, but this would require dynamic redimensioning of matrices and vectors, which could pose problems of its own.

Conversely, a more constructivist approach begins with small layers and gradually supplements them with neurons as the problem / environment demands. In 1990, an architecture exhibiting this progressive, problem-driven growth, known as the cascade-correlation network (Fahlman and Lebiere 1997), actually outperformed contemporary connectionist models on a few benchmarks. Although the idea never really caught on, the same basic principle underlies *boosting* algorithms in machine learning: add new units to handle aspects of a situation that existing units handle poorly (e.g., tricky data cases in a classification task). Furthermore, one of the key contributions of Stanley and Miikkulainen's NEAT system (2002) is *gradual complexification*, wherein networks in the initial population have very few nodes but can then be driven to larger sizes by selection pressure for higher complexity, but only in those problem domains that warrant it; simpler problems are handled by simpler networks.

To fully embrace constructivism (including the selectivity/pruning inherent in Quartz and Sejnowski's full description), an artificial brain needs a developmental phase strongly in tune with its surroundings (i.e., body and environment). Thus, the situations posed externally can exert significant effects on the large-scale structure of that brain, not just on the strengths of its synapses. The debate continues in AI as to whether this level of biological realism offers any significant advantages to engineering design and control (Hiesinger 2021; Miller 2021). The remainder of this chapter continues that investigation via a new PN-POE variant.

7.5.2   Predictive Constructivism

Predictive coding provides an interesting vehicle for POE systems that include artificial constructivism. When development runs in isolation or with restricted environmental contact (as in pre- and neonatal animals), the formation of neural topology receives little guidance from sensory input. Other factors must bias circuit formation. Motor and visceral capabilities surely play a role, as an organism's portfolio of muscles and organs contribute physiologically active destinations for axonal growth prior to any significant environmental coupling. The firing behavior of these peripheral cells and of other internal neurons may serve as targets and/or predictions, and the ensuing prediction errors can then drive learning as the brain builds expectations about its own activity and that of the body.

This contribution of internal drive to early development is emphasized in Buzsaki's (2019) description of body-map formation in young rats:

In the newborn rat pup, every twitch and limb jerk induces a *spindle-shaped* oscillatory pattern in the somatosensory cortex.... The spindle oscillation can serve to bind together neuronal groups that are coactivated in the sensory cortical areas as a result of the simultaneous movement in neighboring agnostic muscles.... Thus, the initially meaningless, action-induced feedback from sensors transduces the spatial layout of the body into temporal spiking relationships among neurons in the brain.

This developmental process is how the brain acquires knowledge of the body it controls or, more appropriately, cooperates with. Thus, a dumb teacher (i.e., the stochastically occurring movement patterns) can increase the brain's smartness about its owner's body landscape.

Thus, when prodded into action by random motor-neuron firing, the genetically determined body structure promotes sensory activity patterns that induce lasting impressions in the brain: the body map. Movement is thus critical to sensory mapping and proper brain development in general.

In addition, general principles from brain evolution and development, such as neural Darwinism (Edelman 1987) and displacement theory (Deacon 1998) could provide important constraints for growing neural networks. Energy considerations may also weigh in, as evidenced by the success of energy-based objective functions in many of the classic network architectures, such as Boltzmann machines, described earlier. In contrast to those networks, which all use a fixed, predefined architecture, other models exploit energy metrics in artificial neural ontogeny.

In their seminal work, Arjen van Ooyen and colleagues (2003) provide considerable empirical and synthetic (i.e., simulated-based) evidence for homeostatic ontogenetic processes that adjust a neuron's collection of incoming excitatory and inhibitory connections in attempts to maintain a particular activity level in the soma. Assuming that those levels are low, particularly for neurons in the upper regions of the hierarchy, then homeostatic development should craft networks to inherently limit neural firing to the bare minimum required for effective information transmission. Those optimal levels surely depend on the organism's sensorimotor demands, which are directly visible to the developing brain via the density of sensory receptors and muscle cells. The relationships between layers in a predictive-coding network perfectly suit this purpose, as each layer tries to regulate activity in the layer below (and hence the amount of bottom-up signals that itself receives) by properly matching top-down expectations to bottom-up reality.

The ability of predictive layers to adapt to one another, independent of global input signals, allows learning and development to co-occur prenatally, prior to the infusion of massive and meaningful sensory data. Thus, the question of what a network tunes itself toward, what worldly patterns it attempts to model prior to significant environmental exposure, is turned on its head: the network adapts to the organism first, the world later.

Predictive networks also enhance the prospects for evolutionary search, that is, the P in POE. The predominantly local interactions of PNs support relatively modular network design without the need for global dependencies. As explained in chapter 6, this facilitates variation by allowing modules to mutate without requiring concomitant changes to other modules. Weakly linked modules can evolve in a much more independent manner than those tightly coupled across broad expanses of time and space, since, among other confounding factors, these distances can introduce signal attenuations and delays that (a) may play significant functional roles due to tight coupling (i.e., proper communication requires a very specific signal protocol), but (b) are easily disrupted by structural change to any of the modules.

7.5.3   The D'Arcy Model

Inspired by D'Arcy Thompson's classic 1942 book, *On Growth and Form* (Thompson 1992), the following model attempts to meld evolution, development, and learning in a manner that will eventually allow researchers to draw clear lines between structure, behavior, and function across the three main levels of adaptivity: evolution, development, and learning. Many ALife researchers (Bongard and Pfeifer 2001; Miller and Banzhaf 2003; Soltoggio, Stanley, and Risi 2018) share this goal, albeit an optimistic one.

The key working assumption behind D'Arcy is that prediction arises from many different neural configurations, so restricting those topologies to one type, such as a Helmholtz architecture or classic predictive-coding layers, will hamstring the search process, thus preventing the discovery of complex heterogeneous structures such as the hippocampus that may indeed prove critical for artificial general intelligence. Thus, the system employs a set of very basic building blocks that admit a host of emergent structures and behaviors.

An artificial genome encodes basic properties of an initial set of neural precursor cells known as *neurites*, each of which is modeled as a central point in two-dimensional space along with two concentric circles, which represent the spatial extent of its axons and dendrites, respectively. Neurites behave similar to those modeled by Ooyen et al. (2003) in that (a) the area of intersection of the axonal and dendritic circles of two neurites determine the (developmental) weights on the inter-neurite links, and thus the intensity of signaling between them; (b) a neurite's accumulated positive and negative signals affect it firing rate; and (c) the deviation of that rate from the desired (equilibrium) firing rate determines the expansion or contraction of the neurite's axonal and dendritic ovals.

Along with evolution, D'Arcy supplements Ooyen's neurite model with local (Hebbian and anti-Hebbian) learning rules such that the total weight on a connection between two neurites (or between the neurons that they eventually produce) includes both an ontogenetic (O) and epigenetic (E) component.

Phylogeny in D'Arcy

At the phylogenetic (P) level, D'Arcy runs a standard genetic algorithm (GA) (Holland 1992; Mitchell and Belew 1996) with a binary genome whose individual genes are divided into N segments, where N represents the maximum number of neurites that a network may contain. Each neurite's genomic segment contains genes encoding the following features (where k is a small integer such as 5):

1. Trigger (1 bit) determines whether or not the neurite will be created at all. Thus, the phenotype network may contain less than N neurites.

2. Influence (1 bit) determines whether the neurite will excite or inhibit other neurites with which it connects.

3. τ (k bits) represents the neurite's time constant.

4. X-coordinate (k bits) of the neurite's location on the 2-d plane.

5. Y-coordinate (k bits) of the neurite's location on the 2-d plane.

6. Axon radius (k bits) of the initial circle of axonal scope.

7. Dendrite radius (k bits) of the initial circle of dendritic scope.

8. Developmental critical period start (k bits) encoding a value in [0,1] that represents the fraction of the neurite's lifetime at which circle growth can begin.

9. Developmental critical period end (k bits) encoding a value in [0,1] that represents the fraction of the neurite's lifetime at which circle growth must end.

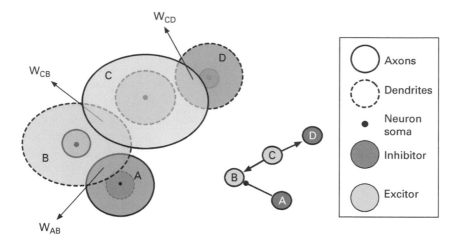

Figure 7.14
(Left) Four neurites (A–D), their concentric axon and dendrite circles, and weights derived from the intersections of one neurite's axon circle with another's dendrite circle. Neurites A and C have more expansive axons than dendrites, while neurites B and D have dominating dendrites. (Middle) Basic topology of neurites A–D showing only the nonzero weights, that is, the weights based on non-empty intersections between axonal and dendritic circles of a neurite pair, with arrow (balled) tips denoting positive (negative) influences.

10. Learning critical period start (k bits) encoding a value in [0,1] that represents the fraction of the neurite's lifetime at which Hebbian learning can begin.

11. Learning critical period end (k bits) encoding a value in [0,1] that represents the fraction of the neurite's lifetime at which Hebbian learning must end.

12. Hebb (1 bit) indicates the learning style (Hebbian or anti-Hebbian) associated with a neurite's dendrites; this pertains to all of a neuron's incoming synapses.

13. Density (k bits, optional) denotes the number of neurons spawned by the neurite.

Ontogeny in D'Arcy
Development begins with a simple translation of the genome's bits into binary flags and real numbers. Those gene segments with a nonzero trigger flag yield a neurite centered at the given Cartesian coordinates, with the encoded radii for its axonal and dendritic circles, a time constant for its dynamics, a learning scheme, and an implicit schedule for when development and learning will begin and end. The critical periods for development and learning may overlap, but learning cannot begin before development begins, nor end before it ends.

Figure 7.14 depicts four neurites, their overlapping process circles, and the weights derived from those intersections. Beginning with neurite A, its axonal radius exceeds its dendritic radius, and the former overlaps with the dendrites of neurite B. Thus, synapses from A to B form, with a developmental weight w_{AB}^d, whose magnitude is proportional to the area of overlap (as scaled by the area of the largest of the two circles, which, in this case, is B's dendritic oval) and whose sign stems from neurite A's genetically determined *influence*, which is inhibitory (-1). Neurite B has very restricted axons: they overlap no dendrites, and thus spawn no synapses out of B. Conversely, neurite C's expansive axonal field intersects the dendrites of both B and D, creating two weights, w_{CB}^d and w_{CD}^d, both of which have a positive ($+1$) sign, since C is genetically excitatory, and both have approximately the

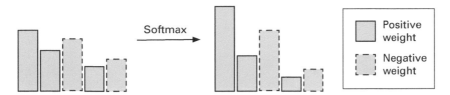

Figure 7.15
Displacement theory in D'Arcy's neurites. Rectangles represent the magnitudes of weights on five incoming connections to a neurite before (left) and after (right) the application of Softmax scaling. This represents a competition for targets in which the most successful invaders tend to increase their advantage, often squeezing out other invaders completely.

same magnitude based on the similar-sized overlapping areas to the northeast and southwest of C's axonal oval.

Equation 7.6 provides the general formula for the developmental weight from neurite i to neurite k ($w_{i,k}^d$), where I_i is the genetically determined influence (excitatory (1) or inhibitory (-1)) of neurite i; A_i^a and A_k^d are the areas of the axonal and dendritic circles of neurites i and k (respectively); and $\bigcap_{i,k}$ is the area of intersection between the axonal circle of i and dendritic circle of k.

$$w_{i,k}^d = I_i \frac{\bigcap_{i,k}}{max(A_i^a, A_k^d)} \tag{7.6}$$

Via a simple supplement to equation 7.6, D'Arcy incorporates Deacon's displacement theory (DT). Recall from earlier in this chapter that DT involves large bundles of axons outcompeting smaller axon groups, due to limited dendritic targets. One common model of competition is the softmax operator (see figure 7.15), which normalizes values in a manner that can (with proper parameter settings) bias the outcome such that *the influential gain more influence* at the expense of the less influential. The competition equation, shown earlier (chapter 4) in the context of the Boltzmann machine is

$$|w_{i,k}^{\tilde{d}}| = \frac{e^{b|w_{i,k}^d|}}{\sum_i e^{b|w_{i,k}^d|}} \tag{7.7}$$

where $|w_{i,k}^d|$ is the original magnitude of the developmental weight from neurite i to neurite k, and b is the biasing factor—larger b entails a higher advantage to the larger magnitudes. Note that all calculations involve magnitudes, with the original weight's sign being reapplied to the scaled magnitude, $|w_{i,k}^{\tilde{d}}|$. In other words, the weight update is

$$w_{i,k}^d \Leftarrow sign(w_{i,k}^d)|w_{i,k}^{\tilde{d}}| = I_i|w_{i,k}^{\tilde{d}}| \tag{7.8}$$

After the triggered neurites with their initial axon and dendrite ovals have been placed on the Cartesian *brain plane*, and all initial developmental weights have been formed (based on all axon-dendrite overlaps), the dynamics of ontogeny can begin for any neurite whose ontogenetic critical period has commenced. All neurites received a randomly generated initial activation level—thus mirroring the tonic firing of real neurons—and the products of these with outgoing synaptic weights yield inputs to neighboring neurites, as in standard neural networks.

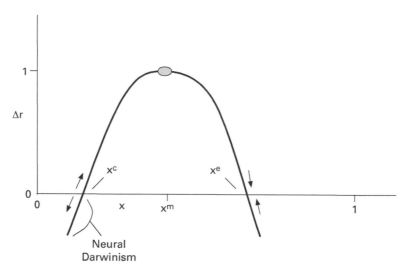

Figure 7.16
The parabolic function for computing changes in a neurite's (axon and dendrite) radii ($\triangle r$) based on its activation level (x). (x^m, 1) = vertex of the parabola; x^c = critical activation level; x^e = equilibrium activation level that feedback dynamics help to maintain: activation levels below (above) x^e cause increased (decreased) radii, which should increase (decrease) activation. Small arrows indicate convergence at x^e and divergence at x^c, both based on the assumption that positive (negative) $\triangle r$ often leads to increased (decreased) x (as discussed in the text).

Neurites abide by CTRNN dynamics, so the sum of weighted inputs is scaled by the time constant τ, and combined with a decayed value of the previous potential (u) to yield the updated potential, which then passes through the activation function (a logistic sigmoid) to produce the current activity level (x).

Next, the key mechanism of Ooyen's neurite model comes into play: the radii (r) of each neurite change as a function of the relative firing rate, a value in [0,1], and the target / equilibrium firing rate (x^e). The relationships among these factors appear in figure 7.16. This parabolic mapping from x to $\triangle r$ is one of several options employed by Ooyen, but it is the main choice in this chapter, because it provides a straightforward abstraction of neural Darwinism's *survival of the best networkers* via the leftmost strand of the curve, which dips under the *x*-axis. Hence, when activation slides below the critical value (x^c), the radii shrink. This will often initiate a negative ontogenetic spiral for the given neurite, since reduced radii entails less contact with other neurites and hence less signal input. One caveat is that incoming signals can be both excitatory and inhibitory, so a reduction in contact with inhibitors can enhance firing activity. However, if a D'Arcy run converges on *biological conditions*, such as the (approximately) 5:1 ratio of excitatory to inhibitory units (Kandel, Schwartz, and Jessell 2000), then most neurites with low activation levels will suffer the miseries of neural selection as their radii and firing rates descend toward zero.

Parabolas such as those in figure 7.16 that have a vertex at (x^m, 1) and open downward have the functional form of equation 7.9, which is abbreviated as $G_d(x)$ to denote the developmental growth function of a neurite's activity level (x).

$$G_d(x) = \frac{-4}{(x^e - x^c)^2}(x - x^m) + 1 \tag{7.9}$$

To incorporate critical periods (explained below), the phase of a neurite's developmental critical period ρ^d factors into equation 7.9 to yield the complete radial growth function of equation 7.10, which then applies to both the axonal and dendritic ovals of neurite k.

$$\triangle r_k = \rho_k^d G_d(x_k) \qquad (7.10)$$

This relationship ensures that a neurite's processes will only grow or shrink when (a) the neurite is within its developmental critical period, and (b) its activation deviates from x^c and x^e. A stable neurite structure is therefore one that has either lost developmental plasticity or can consistently maintain an activation level near the target / equilibrium value, x^e.

Epigenesis in D'Arcy
Basic Hebbian principles govern learning in D'Arcy, although these can easily be replaced with more elaborate (but still local) synaptic update rules. The learning component of a weight from neurite i to k, $w_{i,k}^l$, is given by equation 7.11, where ρ_k^l marks the phase of the learning critical period, λ is the learning rate, h_k is the genetically determined Hebbian style (normal $(+1)$ or anti-Hebbian (-1)) of the postsynaptic neurite, and the x's are activation levels. All learning weights are restricted to nonnegative values, whereas the developmental weights can take on positive or negative values depending on the excitatory or inhibitory character of the pre-synaptic neurite.

$$\triangle w_{i,k}^l = \lambda \rho_k^l h_k x_i x_k \qquad (7.11)$$

To account for both the developmental and learning effects in an integrated manner, thus partially blurring the distinction between the two adaptive processes, D'Arcy models connection weights as the sign-adjusted geometric mean of the learning- and developmental-weight magnitudes, as expressed by equation 7.12.

$$w_{i,k} = sign(w_{i,k}^d)\sqrt{|w_{i,k}^d||w_{i,k}^l|} \qquad (7.12)$$

Notice that the geometric (as opposed to the arithmetic) mean ensures that if either component vanishes, so too will their combination. Hence, in cases of small $|w_{i,k}^d|$, learning will struggle to produce a strong influence of neurite i on neurite k, thus representing a topology in which there are simply not enough i-k synapses to strengthen. Conversely, a high overlap between i's axonal and k's dendritic ovals will ensure high *bang for the buck* of any Hebbian weight changes.

The hybrid weight scheme also allows development and learning to proceed at their genetically predisposed paces (via the evolved critical periods), with either little or considerable overlap between the two.

Critical periods are also modeled as downward-facing parabolas (see figure 7.17), representing functions of time, with a maximum ρ value of 1, signifying full effect. Evolution determines the two x intercepts (within the segment [0,1] of the time axis) for a neurite's two parabolas: one each for development and learning. Time values within [0,1] denote fractions of the agent's lifetime.

7.5.4   Neurites to Neurons in D'Arcy

A key motivation behind the DEACANN system (Downing 2007a) (discussed above) was to abstract development to a computationally feasible level (on a laptop). Thus, axonal and

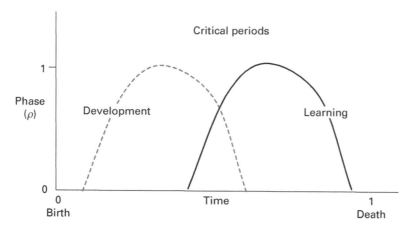

Figure 7.17
Overlapping parabolas representing the phases of critical periods for both development (ρ^d) and learning (ρ^l). Each neurite has its own two parabolas, with x intercepts encoded in the genome.

dendritic tags were introduced to circumvent the detailed simulation of axonal and dendritic growth for hundreds or thousands of neurons. However, DEACANN handles the two adaptive processes in lockstep: development must end before learning begins. D'Arcy employs a different abstraction strategy, and one that more seamlessly integrates development and learning.

The D'Arcy model presented above assumes a simple equivalence of neurites and neurons: each neurite represents a single neuron. Thus, it does model development at a relatively detailed level in terms of the extension and retraction of the axons and dendrites of individual neurons. However, as a simple enhancement, the genome can include a *density* gene (listed earlier as optional) to denote neuron cardinality. Each neurite then represents a collection of neurons, with development still simulated at the neurite level but learning now proceeding at the finer-resolution interneuron level.

For development, the neurite's activity level is simply the average activity over its neurons, and this then determines $\triangle r$, just as described earlier. Learning involves paired neuron activation levels and employs slight variations of equations 7.11 and 7.12, with the i's and k's associated with general neurite properties (or specifically with development) replaced with n(i)'s and n(k)'s to signify the neurites to which neurons i and k belong. Equations 7.13 and 7.14 handle these extensions.

$$\triangle w_{i,k}^l = \lambda \rho_{n(k)}^l h_{n(k)} x_i x_k \tag{7.13}$$

$$w_{i,k} = sign(w_{n(i),n(k)}^d)\sqrt{|w_{n(i),n(k)}^d||w_{i,k}^l|} \tag{7.14}$$

This scheme separates development and learning so that the former can run at a coarser spatial resolution but still in parallel with learning (when their critical periods overlap). Thus, perturbations due to internal noise or sensory input can influence both development and learning, depending on the neurites, neurons, and critical periods. In this way, D'Arcy facilitates constructivism: the environment can affect neural development.

7.5.5   Peripherals in D'Arcy

The interface between a D'Arcy network and an agent involves special sensor and actuator neurites, with the former (latter) having only an axonal (dendritic) oval. These circles abide by the same activity-based developmental rules as those of other neurites. Sensory input to the agent then directly determines firing activity of sensory neurites, and the activity of actuator neurites translates into motor settings of the agent. The genome encodes most of the same basic features of sensor and actuator neurites, such as location, initial radius, learning style (for motor neurites), and so on.

When ALife experiments include evolutionary search for useful morphologies, including the locations of an agent's sensors and actuators, the evolved Cartesian coordinates for their corresponding D'Arcy neurites can be easily mapped to points on the agent's body. This is particularly relevant for sensors, many of which reside in the head (i.e., near the brain), but it also pertains to actuators directly involved in sensing, such as the emitters of sonar signals found in the mouths of echolocating bats. When, and at what frequency, to emit a sonar signal is an important decision for an echolocator (Dusenbery 1992), and the relative locations of emitters and echo receivers can also greatly impact performance. Adaptivity of behavior and morphology go hand in hand, and D'Arcy provides basic support for both.

7.5.6   Neuromodulators in D'Arcy

As championed by numerous researchers (Houk, Davis, and Beiser 1995; Doya 1999; Edelman and Tononi 2000; Soltoggio et al. 2008), value systems are integral to the successful behavior of autonomous agents, primarily as feedback signals for reinforcement learning (Sutton and Barto 2018). Thus, all correlations between the firing rates of paired neurons should not have equal effects on learning: only those occurring during periods of high reward or penalty (i.e., positive and negative value signals) should significantly influence synaptic change, as described earlier (chapter 3) in the context of the basal ganglia.

Unlike synaptic information exchange from one neuron to the next, value signals in the brain typically involve neuromodulators (dopamine, acetylcholine, and so on) secreted into extracellular spaces and thereby influencing thousands of dendritic trees simultaneously, with the net effect being the promotion of learning on those synapses that both (a) experience coincident pre- and post-synaptic firing, and (b) feed into dendrites that express receptors for the particular neuromodulator (Hasselmo, Wyble, and Fransen 2003; Prescott, Gurney, and Redgrave 2003). Thus, neuromodulated learning constitutes three-way coincidence detection of the value signal, pre- and post-synaptic firing.

The axonal and dendritic ovals of D'Arcy's neurites provide an ideal backdrop for neuromodulator modeling. Assuming that certain neurites are designated as value centers, then a center's axon circle circumscribes its zone of influence: all other neurites with dendritic ovals that intersect that zone will be susceptible to modulation, with the degree of influence correlated with the area of overlap. The collective activity of all neurons in the value neurite then determine the amount of neuromodulator secreted into the adjoining dendritic ovals, each of which then applies the same modification factor to all local learning in all neurons of the receiving neurite.

The generic learning rule, updated to include neuromodulation, then appears in equation 7.15, where the new term, $m_{n(k)}$, reflects the total amount of neuromodulator (from all overlapping value centers) received by the dendrites of neuron k's neurite.

$$\triangle w_{i,k}^{l} = \lambda \rho_{n(k)}^{l} h_{n(k)} m_{n(k)} x_i x_k \tag{7.15}$$

Value centers should otherwise function as normal neurites, with dendritic trees that determine the influence of upstream neurites on value judgements. Alternatively, a value center can be declared as a sensory neurite, with all stimulation coming from outside the network.

7.5.7 Predictively Unpredictable

The above details form the core of the D'Arcy framework, while many of the other devilish details are still open to modification.[3] By offering evolution such low-level primitives, which will then determine many aspects of development and learning, D'Arcy creates an immense search space with all of the concomitant challenges. However, the wide variety of predictive mechanisms and networks covered in this book should serve as ample justification for a very open-ended search for fruitful neural morphologies and dynamics. Hopefully, future research will confirm this optimism.

7.6   Most Useful and Excellent Designs

The enormous success of deep learning (DL) has (mis)led many to believe that AI *is* ML, which *is* DL. Naturally, this is a sour pill to swallow for an old-timer who has revered AI pioneers, such as Nils Nilsson (1980), Patrick Hayes (1979), and Allan Newell and Herbert Simon (1972), and learned to love planning algorithms, nonmonotonic logics, and an arboretum of tree-search methods. But who can argue with good results on real-world data, as opposed to perfect results on toy problems? When inquiring about one of my AI classes, a geology graduate student politely informed me that *AI is just an excuse for computer scientists to do some statistics*. Ouch!

But when a pioneer such as Geoffrey Hinton begins to question the ultimate potential of conventional backpropagation grounded in long-distance gradients, that has to raise some eyebrows. One group of researchers who might hope to exploit this chink in the DL armor is those who evolve the weights of neural networks as their sole adaptive mechanism. As discussed above, these teams can leverage the same hardware advances (GPUs, TPUs, and so on) as do DL aficionados, thus bringing evolutionary-algorithm (EA) populations of thousands of individuals, each containing millions of weight encodings, well within the realm of possibility, and obviating backpropagation completely for certain tasks, such as finding good neural-net-based controllers for complex processes. Alternatively, EA experts seeking a more cooperative relationship with DL may evolve network architectures but leave the weight tuning to backprop, as done by NAS adherents.

Prediction provides yet another scenario. As discussed above and detailed in chapter 5, several researchers (Whittington and Bogacz 2017, 2019; Song et al. 2020) promote predictive coding as a viable alternative to long-distance gradients; their learning systems achieve very competitive classification results while performing many of the same basic calculations as backprop in a local, biologically compatible manner. Similarly, the two-phase learning employed by NGRADs (see above) such as the Boltzmann machine provide Hebbian-based alternatives that Hinton and other DL luminaries still view in a very positive light. Chapter 4 on energy networks hopefully provides more than mere historical context for modern DL. Those techniques, most of which embody prediction and learning driven by prediction error, could easily return in future AI . . . whoops ML . . . ah . . . er . . . DL cycles.

Is that as far as this might go, then? Will predictive-coding networks of a few classic forms be our best shot at incorporating this essential feature of cognition (ubiquitous prediction) into artificial neural networks? This chapter argues for an even deeper plunge into biology, because as shown in earlier discussions of neural regions such as the hippocampus, neocortex, and basal ganglia, predictive networks exhibit a wide variety of topologies. We can only speculate as to which among these (and many other options) offers the best opportunities for artificial general intelligence (AGI). But for AI search algorithms to find those glowing zones of design space, they must be equipped with a set of primitives that strike that ever-elusive sweet spot between biological nuance and oversimplified computational abstraction. I believe that those primitives come from this book's basic breakdown of prediction and its parallels to control: excitors, inhibitors, delays, derivatives, integrators, comparators, local learning rules, and varying time constants. CTRNNs seem to provide a minimally complex neural substrate for realizing all of these features, either directly, indirectly, or with a few minor twists (such as restricting units to either excite or inhibit all neighbors).

To fuel this extensive search, modern-day computing power can be harnessed for yet another chore: the simulated development of neural networks that are perpetually in tune with their environment, as constructivism advocates. The same computational advantages that elevated DL to AI's golden child, and that allow weight-encoding EAs to keep pace, can also pump a fresh wind into the sails of the many POE algorithms that so rarely make it beyond the dry dock. They can actually emulate the evolution, growth, and fine-tuning of designs that could not be found otherwise.

On the final page of *On Growth and Form*, D'Arcy Thompson wrote, "Moreover, the perfection of mathematical beauty is such . . . that whatsoever is most beautiful and regular is also found to be most useful and excellent."

This succinctly and poetically summarizes his main argument: the *useful and excellent* structures of living organisms are not the sole products of random variation and natural selection but owe a significant debt to the forces of physics and chemistry (in all of their mathematical glory) playing out over developmental time. His book is one long, eloquent, and example-rich argument for the role of physical forces in crafting organic structures. The behaviors produced by those structures then serve various functions (aka purposes or teleologies), and the trio of structure, behavior, and function becomes the composite grist for natural selection. Thompson largely restricts his analysis to structure / form and its developmental emergence, but when evolving brains for bots in POE systems, the entire spectrum, from growth to purpose, comes into play.

This presents a much more challenging problem of linking growth, via structure and behavior, to purpose. It also forces a reckoning of biology and computation, since our technologies can exhibit similar behaviors for similar purposes as natural systems, but via completely different structures. Hence, the natural processes for growing those structures may be wholly irrelevant for the corresponding engineering solution. This is one of the key problems with many synthetic EvoDevo systems: the simulated natural growth routines require too many constraints and hacks to produce useful tools. Most POE systems are relegated to a dustbin full of eclectic *proofs of concept*.

Prediction affords an interesting vantage point and potential solution to this problem. If, in fact, prediction is a fundamental function of the brain, and if the relationships between fine-grained subdivisions (e.g., cortical layers, motor ganglia, and so on) are also predictive, then this decomposition of global function to similar local functions manifests a *functional fractalization* that developmental growth processes can leverage to *presciently adapt*, prior to any actual experience in the world.

During many discussions among colleagues concerning development and its potential utility in AI, we often ponder the role of environmental exposure during early growth. The analogy to an embryo in the womb frequently comes up. It receives preprocessed nourishment and coarse-grained environmental signals (such as muffled sounds and damped vibrations), and its internal structures then change in response to that information in concert with its genetic recipe. The question then arises as to whether, during development, a POE AI system could benefit by receiving reduced, abstract versions of the problems that it will tackle as a mature agent. For example, a developing classifier net might receive only grainy, low-resolution images, or time series in which each value is an average over a large temporal window. The problem, of course, is that a classic DL classifier needs to be fully configured to do any adaptation, since the long-distance gradients require complete paths (forward and backward) connecting parameters to outputs. Hence, the predigested data approach seems futile: once the net is fully configured, why give it anything but the real, high-resolution data?

However, if we now adopt the inside-out neural perspective of Buzsaki, Llinas, Clark, and others, then external signals no longer constitute a driving factor for behavior (including learning). Internal drives become primary. These include the maintenance of reasonable firing rates and the efficient transmission of relevant information with minimal energy; and these lead to cooperation and competition among neural groups, which, in turn, induce local predictive topologies in which one group can quell the activity of another by anticipating its patterns and neutralizing them at a comparator.

Functional fractalization allows the growth processes to get a good taste of their intended function, prediction-making, very early in the agent's life, thus allowing local subdivisions to refine their skills long before exposure to the *whole problem*: the real environment. Prediction at the macro level, in terms of an agent's overt ability to foresee future events, provides a nice challenge for many AI systems. But it is at the micro level that prediction may offer its greatest contribution, by providing local tasks trained with local feedback (i.e., prediction errors). This chapter gives some indication of how ALife POE systems can leverage this functional fractalization in the quest for emergent general intelligence.

8 Conclusion

8.1 Schrodinger's Frozen Duck

Turning left off of E6, Norway's main interstate highway (a two-laner with a decent shoulder), we begin the final 30 kilometers (on a two-laner with no shoulder and sporadic midline markers) of our yearly Christmas trek to Mormor's (Grandma's) house. At this exact point in the journey, the excitement mounts, not so much for the anticipation of presents under the tree, but for the start of yet another ridiculous family contest: predicting the temperature at Mormor's house, located in one of Norway's coldest towns, Folldal.

At the turnoff, my wife and I read aloud the current outdoor temperature, often around $-15C$; then everyone has to pick a unique integer. The backseat is alive with calculational cacophony as each child promotes their number with a geographic/meteorologic justification. There is little elevation change from the turnoff to Folldal center, but from there up to Mormor's house is a steep climb with anywhere from a 5 to 15 degree temperature swing, which is often positive, since cold air sinks.

The arguments rage for a kilometer or two before we force everyone to lock in their prediction. There are the occasional, illicit attempts to text ahead to Mormor to get a mercury reading, but the kids tend to self-police such blatant cheating attempts. Elevated vigilance levels pervade the tense atmosphere as the minute-by-minute temperature readings allow everyone to evaluate their winning prospects. Gradients of temperature change per kilometer also enter the picture: *The temperature dropped 3 degrees since the turnoff, and we're now halfway to Folldal, so . . .*

Nothing is decided until we actually pull in the driveway and park the car, since the temperature has been known to change several degrees in the last 100 meters of the trip, during the final ascent. The winner gets bragging rights for the frigid holiday season, of which they tend to take full advantage.

Wrapped inside this half-hour contest are a host of smaller-scale predictive acts, such as the numerous expectations that the driver employs to safely navigate treacherous backroads across a frozen landscape. Flashes of light when entering sharp turns reveal oncoming vehicles, while reflections from the pavement provide invaluable warnings of upcoming black ice, and odd roadside background shifts may be the only clue one gets of a crossing moose. In the backseat, any suspicious roll toward the door may be a dead giveaway of cell-phone cheating.

Beneath these moment-by-moment predictions are split-second brain activities involving expectations, some explicit (conscious and verbalized) and others merely implicit, but all supporting the decision making of the competitors and driver. And all involve predictions more about future brain states than about future states of the world as a whole, since, for the most part, the temperature at Mormor's when we reach the turnoff is the same as when we reach her driveway. It's our awareness of that reading that lies in the future, and for the sake of the contest, that is all that matters.

This subjective nature of *future* becomes evident in another consequence of my geographical situation: delayed awareness of American sporting outcomes. As an avid fan of the University of Oregon Ducks, I am frequently plagued by the nine-hour time difference between Scandinavia and Oregon. I retire on Saturday evenings in heightened anticipation of an Oregon football or basketball game that will start and end during my slumber, and my dreams often include wild predictions of win and loss scenarios. Could the team mascot score the winning points? Could the pounding Oregon rain suddenly freeze and turn a football game into a hockey match? I awaken on Sunday morning into *Schrodinger moments*: until I open my digital tablet, the ducks are both victorious and defeated *where it counts*, in my mind.

Of course, my state of mind means nothing to Las Vegas bookies; when the final whistle blows on the field or court, speculations of future outcomes become objective exchanges of cold hard cash, and the case closes for all intents and purposes. In contrast, I hope that this book has given readers an appreciation for prediction as a very personal internal phenomenon, one that is rich and expansive in scope, yet grounded in simple, ubiquitous, neural mechanisms.

This book began with that famous quote by Yogi Berra: It's tough to make predictions, *especially about the future.* After a few hundred pages, the oxymoron in that statement may have lost some vigor. When expectations involve projections about a subjective information state concerning present and past world states, the bond between prediction and future attenuates. Through that weakening, the concept expands in scope, which I hope has encouraged readers to see the broader perspective of prediction.

8.2   Expectations Great and Small

Underlying any overt prediction, such as that of tomorrow's soybean futures, run a host of declarative and procedural expectations, many of events as simple as the upcoming activation levels of particular neurons. At the end of chapter 7, I used the term *functional fractalization* to highlight the similarity of the overt and low-level functions / purposes: they are all predictions. However, akin with most functional decompositions, they clearly involve different mechanisms and scales.. Although a stack of cortical columns presents a nice image of modular predictive coders, and possibly a productive computational model for certain tasks, it omits many key differences between peripheral and internal neural processing. These discrepancies are vital to a neuroscientist but anathema to AI researchers looking for a short list of basic principles that can explain as much of intelligence as possible. So those of us in sciences of the artificial will jump at opportunities to flatten our learning curves with a juicy serving of the finest abstractions. Prediction is one such meal.

This need not paint a reductionist picture of intelligence as *prediction all the way down*. Rather, the moral of this and other books (Llinas 2001; Clark 2016; Buzsaki 2019) might be closer to this:

Everything is not prediction, but prediction is everywhere.

The predictive primitives described in these chapters (delays, gradients, averages, and so on) are ubiquitous in neural networks (both natural and artificial), but they do not always combine and interact in support of prediction. Still, their omnipresence within a diverse collection of neural circuits makes the emergence of predictive motifs nearly unavoidable. Facilitated variation, development, and learning provide means, and selection provides myriad motives.

Proficient predictors surely have a selective advantage over organisms whose mental world is confined to the present. A good deal of intelligent behavior involves supplementing the present with enough relevant memories of the past to imagine the future. Of course, fully knowing the future is impossible, but having biases that correlate well with it can only increase one's odds of survival. And, as detailed earlier, the temporal disparities between fast motor actions and slow sensory processing in most species rewards any mechanisms that can provide reasonably accurate hints as to near-future sensor readings. At the cellular level, basic constraints such as energy and information efficiency would favor predictive-coding schemes that can reduce neural firing as much as possible while still maintaining the essential information coupling needed to keep a body running smoothly, both inside and out.

As overviews of the cerebellum, hippocampus, basal ganglia, and neocortex should indicate, predictive circuits can take many forms; and the function of expectation generation can peacefully coexist and cooperatively dovetail with vital faculties such as memory, perception, and action selection. Unfortunately, each of these neural structures exhibits a level of cellular heterogeneity and behavioral complexity that precludes any quest for basic principles of neural prediction. These are hardly the frictionless planes of neuroscience. However, the myriad species- (or class-) specific predictive topologies further highlight a selective pressure to develop something (anything) that can bring an organism's visions of tomorrow a little closer to today.

8.3   As Expected

The search for general predictive principles may also lead deep into development, where cooperative and competitive interactions among neurons and neural groups (fortified by Hebbian STDP) may have led to the (nearly inevitable) emergence of predictive motifs consisting of top-down signals, delays, inhibitors, comparators, and bottom-up error signals. In short, prediction and control may have arisen as natural consequences of facilitated variation's modularity and weak linkage (Kirschner and Gerhart 2005), neural Darwinism's *survival of the best networkers* (Edelman 1987), and displacement theory (Deacon 1998).

Since the natural world as *we* experience it puts a premium on predictive competence, the fact that evolution found respectable forecasting techniques should come as no surprise. However, a sloth lives under different constraints than a human; its actions are very slow. One can imagine many worlds, based on carbon or silicon, where sensing and acting have more similar timescales—or where vision, working, after all, with the fastest known entity (light), fully eclipses motricity—and thus the obvious need for prediction diminishes. Of course, without fiber-optic brains, we still need to differentiate the speed of waves and particles from the speed at which our default machinery can interpret them. Robots may circumvent this problem: a well-designed humanoid may have very high-frequency sensing but slower motricity, particularly if it should safely interact with humans. Thus, from the sensorimotor perspective, a robot may have less need for prediction than a human.

However, any truly intelligent agent, real or artificial, requires lookahead: the ability to envision and evaluate future options, only some of which will actually come to pass. Unless it can physically backtrack (i.e., run) at the speed of light, an agent must occasionally commit to irreversible choices whose consequences will be known only in the future. For example, a chieftain preparing to visit an unfamiliar new tribe may have to choose between spears and body armor, or gifts and festive garb, with serious ramifications for either mismatch. The contents of a fisherman's boat as he heads out to sea can have extreme repercussions later in the day, with no second chances for repacking. Carefully designed experiments show that even ravens can plan ahead by choosing the proper tool for a task that they will encounter minutes or hours in the future (Kabadayi and Osvath 2017). A good deal of higher intelligence would simply not be possible without prediction.

In *Intelligence Emerging*, I wrote a lot about search and its essential contribution to the emergence of intelligence, across multiple spatiotemporal scales. Predictive machinery is one of the golden nuggets that all that search eventually found. It props up very high points in the fitness landscape. A revised theory of facilitated variation (Kirschner and Gerhart 2005) might one day include predictive coders in its list of *core components* found by evolution and then repeatedly exploited to ratchet up complexity and intelligence.

Thus, it seems that predictive machinery has eventually emerged, *as expected*, given the world in which we live and evolve. Any crystal-ball mystique surrounding prediction has hopefully been dispelled by these chapters, which both broaden the range of predictive activity and also give indications of how the overt predictions of our daily lives have reasonable neuroscientific explanations, many of which involve the brain's own version of forecasting. Expectations are a natural part of our evolutionary past and present, and the future prospects for automated prediction seem bright.

8.4　Gradient Expectations

The recent, wild successes of deep learning (DL) have created overwhelming expectations for both itself and AI as a whole. Though many of the predictions, such as fully autonomous automobiles, have had *reality checks* as researchers understand the difficulties of the remaining (peripheral but essential) aspects of the problem (i.e., recognizing pedestrian intentions), these have not derailed DL. There are too many triumphs to simply discard these techniques as *science fiction*[1] and scurry back to the safety of more traditional science and engineering approaches, those whose results can be verified mathematically, or empirically, in the lab.

As of around 2012, DL has proven that gradient-based methods perform expertly in a wide range of domains, when given enough data and computing power, both of which have become abundant in the past decade. Long-distance gradients have withstood stellar challenges from the population-based search methods of evolutionary computation, and they have partnered with the trial-and-error search techniques of reinforcement learning (RL) to form the DRL colossus, which has brought the world's chess and go champions to their knees—and all in the absence of expert domain knowledge and human-game data cases. DRL success now requires only a good programmer and powerful machines. DRL workers have masterfully weaned themselves from the data-as-oil that many companies have leveraged to attract AI partners. For better or worse, AI progress can continue quite independently in those areas where the basic principles and rules are freely accessible (e.g., in textbooks) and available computation permits extensive self-investigation / self-play by a digital agent.

The looming question is whether DL and its gradients can get us all the way to artificial general intelligence (AGI). Skepticism abounds (Mitchell 2019; Larson 2021; Hawkins 2021), as does the push for a return to biology for more hints and inspiration (Hiesinger 2021; Soltoggio, Stanley, and Risi 2018; Miller 2021). At the same time, nascent Hebbian learning methods, based on predictive coding (Whittington and Bogacz 2017), can preserve many of the powers of backpropagation. These local mechanisms, along with some of the other predictive circuitry discussed in this book, could carry neural networks beyond the jumbo-gradient era and into a more versatile, biologically realistic future.

A broader philosophical view of AGI and its origins further weakens arguments about the supreme importance of DL and super-sized gradients. A common artificial life (ALife) perspective on AGI is that it can only arise via the interactions of agents with other agents and their environment over time periods spanning many generations. As (most) early AI researchers eventually realized, you cannot simply pound AGI into a machine as logical rules for understanding and behaving in the world; and as (some) contemporary DL adherents would probably admit, you cannot feed millions of examples into a neural network and expect it to induce all salient generalities from those images, text, sounds, and the like. The value of actually exploring the world and generating data oneself has been grossly underestimated by much of AI, although self-play in DRL systems such as AlphaZero has surely opened many eyes and minds to the possibilities and advantages.

As discussed earlier, prediction is a nice trick for creating personal data sets for training ML systems to generate expectations from current states. Thus, supervised learning can surely play an important role in AGI. However, our brains probably do a lot more unsupervised and reinforced learning than supervised, as implied by the proposed functional breakdowns of various brain regions, such as neocortex, hippocampus, and cerebellum, where only the latter exhibits anything close to truly supervised learning (Doya 1999). We cannot employ semantic deceptions such as calling DL's autoencoders unsupervised learning just because the input and target are the same; the training algorithm is still backpropagation. Of course, this book argues for a different semantic stretch: viewing one level as producing a target value for the predictions of a neighbor layer, with the ensuing prediction error providing all the feedback needed for effective learning. Neither can the billions of local gradients in this extensive population of PID controllers pave the whole road to AGI on its own.

The expectations generated by gradient-based methods seem overblown and unrealistic to anyone who takes seriously the connections between nature and engineering. As I argued extensively in *Intelligence Emerging*, the role of persistent but relatively random search in all phases of evolution, development, and learning seem absolutely fundamental to the design of cognitive machinery. Readers of this book have hopefully gained an appreciation for a wider array of gradients, how they enable prediction, and how the mechanisms for handling local gradients fit nicely into various accounts of the evolution of intelligence. In this way, local gradients and prediction constitute important bio-inspirations for the continuing pursuit of AGI.

8.5   Expecting the Unexpected

Despite many years in the field, I am probably no better a predictor of AI's future than, say, Yogi Berra. If my technological forecasts over the years had been wagers, I would surely be penniless. The predilections formed by several decades of deep technical immersion (in the neurons of the brains of the bugs in the trees) often cloud one's view of the forest. My only antidote to this myopia is reading higher-level accounts written for the general public; and in the past decade, a flood of such books, on AI, have hit the market.

Some paint a very grim picture of a denuded planet ruled by robots, with us as their slaves. Others strike a more reasonable balance of power in which the human-machine cooperative reaches highly productive and socially and emotionally pleasing levels. It's hard to avoid a sense of AI awe when reading a nearly perfect DL translation of English to French; or viewing microscopic images of tumors that machines, but not humans, could detect; or watching a robot hurdle obstacles like an Olympic champion. But, giving pause to even the most optimistic futurist are tales of racist bots that actually know nothing of race or suffering, medical ML systems having no understanding of basic biology, image classifiers thrown off by a little piece of tape on a stop sign, and "autonomous" robots controlled by an engineer hidden behind a shopping-mall palm tree. I am often tempted to view the current (very hot) AI summer as little more than prelude to another AI winter, as the roller-coaster of hype, hope, and disillusionment rumbles on.

Ray Kurzweil (2012, 2005) is one futurist whose visions resonate well with my own interests and experiences. One of his key speculations concerning the *singularity* (a term coined by Vernor Vinge in the 1980s) is an accelerating feedback relationship between AI and neuroscience. As AI, robotics, and hardware improve, so does our ability to measure and interpret neural data, and thus our ability to understand the mind. This enhanced comprehension of the brain then cycles back to fortify our AI systems. As the hardware and algorithms improve, so too does the frequency of this cycle, eventually leading to a state in which the AI bots are intelligent enough to design their own neural theories, and the robots have the dexterity to physically test them. Humans are then politely ushered out of the loop, and into the cheap seats in the back of the arena, and progress proceeds at lightning speeds, yielding systems that dwarf human skill and intelligence. Stop the level-5 autonomous Lamborghini; I want to get out and walk!

Putting dystopian drama aside, the positive feedback between AI and neuroscience at the core of Kurzweil's theory has very positive overtones of mutual progress for both fields. Although the ratio of intersection to union of these two disciplines is small, there are a sufficient number of researchers worldwide with the cross-disciplinary fortitude to investigate intelligence from both the natural and artificial perspectives. Still, a great many pilgrimages by AI workers to nature's holy ground result in little satisfaction, spiritual or otherwise. The bug of biological inspiration bites many, but the victim grows weak of trying to turn the fascination into a competitive algorithm. But just as we continue to trawl the vast rainforests for miracle cures, there is no reason to discontinue our trips across campus to the neuroscience department.

AI's involvement in another, quite different, positive feedback has decidedly negative consequences: the viral spread of outrageous hyperboles, vicious insults, and bald-faced lies. We live in very unnerving times, when democracies have begun a spiraling decline fueled by ever-growing fears and animosities, for which AI has played no small part. So many of these incredulous fantasies and conspiratorial chimera tear through layers of cyberspace unhindered, never meeting reality at sorely needed comparators, but instead feeding back into and magnified by the overheated matrix. It is imperative that humanity, in tandem with our technologies, actively addresses the dangers that have been so widely and loudly anticipated, before all red lights are on, and all bets are off.

The simple act of juxtaposing fiction (no matter how innocent) with fact, identifying the differences, and using them to improve the quality of our collective reporting (to legions of followers) would go a long way. Error and failure are, after all, the catalysts of learning and improvement, but only when recognized and admitted. In *Great Expectations*, Dickens (1861) addresses this adaptivity that often emerges from hardship:

I have been bent and broken, but—I hope—into a better shape.

Unfortunately, Yogi Berra paints a different picture:

The future ain't what it used to be.

It may be just as enlightening to let AI predict its own future. When given the first five italicized words, DeepAI's text-generation system completes the thought (and artfully segues into acknowledgments):

In the future, artificial intelligence will be able to learn what you thought you knew, but humans will just have to figure out the difference from what you actually had. I want to thank the team for working so hard to find a way to make this possible and thank my backers for being involved.

Whether the classics, the comical, or the artificial should serve as our guide to the future is anyone's guess. But most readings of the technological gradients indicate that AI will play an oversized role. Ours is to pay extremely close attention.

Notes

1. Introduction

1. In baseball, a strikeout occurs when a batter misses on three attempts at hitting a pitched ball, while a round-tripper is a slang expression for a homerun, in which the batter hits the ball (typically very far) and is able to run around all bases and back to the starting base (home) to score one or more points, depending on the number of teammates who were on the bases at the time.

2. I experience this temporal mismatch when standing in an amusement park's baseball batting cage in which the velocity of incoming balls has been set very high. The approaching ball is visible, albeit blurry, but I cannot even begin to react before it strikes the backstop. My body freezes, totally dumbfounded by it all.

2. Conceptual Foundations of Prediction

1. The more common term is *allocentric*, but its standard connotation involves the perspective of other people, while its philosophical and scientific meaning is *from the perspective of anything other than the individual.*

2. It is hard to read a DL article without encountering multiple occurrences of the word *gradient*.

3. The direction of people's mental number line tends to mirror the direction in which they read. Hence, most native English speakers view larger numbers to the right, while Arabic speakers envision them on the left (Dehaene 1997).

4. Notice that this is actually three nested averages: (1) Each student's GPA is the average grade over all of their courses. (2) These student' GPAs are averaged to create the mathematics department's average for its incoming student group for a particular year. (3) The year-group averages are averaged to make a prediction. The first two averages are normally unbiased, while the third may use unequal weights to improve predictive accuracy.

5. It seems reasonable to assume that $k_g + k_a = 1$, but forgoing this constraint helps highlight the similarities between prediction and control.

6. Point guard and forward are roles performed by basketball players.

3. Biological Foundations of Prediction

1. In general, sophisticated motor activity demands widespread inhibition. In fact, local actions often require more restraint and thus constitute more complex activities, as mentioned earlier about moving one versus all fingers or toes.

4. Neural Energy Networks

1. Hopfield's original neurons used 0 and 1 but essentially scaled them to -1 or 1 at the start of most calculations.

2. In the original nomenclature, an RBM consists of one pair of layers, while the full network comprises a stack of RBMs, with one variant known as a deep belief network (DBN). However, other literature uses RBM more generically, as either one or many paired layers. This book uses the generic connotation of RBM.

3. The term $p_g(s|d)$ is an odd conditional probability, since the generative weights produce data (d) from internal states (s), not s from d. Simple use of Bayes rule expresses this in terms of three probabilities that make more intuitive sense: $p_g(s|d) = \frac{p_g(d|s)p_g(s)}{p_g(d)}$. Here, $p_g(s)$ are just the a priori internal-state probabilities.

4. This would make a nice motto for an alternative-living commune.

5. Predictive Coding

1. Of course, the value of $F'(s_k)$ depends on the activation function, F. When F is linear, $F'(S_k) = 1$, while the logistic activation function (aka sigmoid) has $F'(S_k) = F(S_k)(1 - F(S_k))$, and the hyperbolic tangent has $F'(S_k) = 1 - F(S_k)^2$.

2. The negative signs in these rules assume an objective function of positive magnitude (e.g., free energy) that predictive coding strives to minimize. Alternatively, Whittington and Bogacz (2017) maximize a negative energy function and thus have no negative signs in their learning rules.

3. The brain accounts for roughly 2 percent of the body's mass but uses about 20 percent of its energy (Sterling and Laughlin 2015).

6. Emergence of Predictive Networks

1. The expression *ontogeny recapitulates phylogeny* implies that key evolutionary transitions in the ancestral lineage of a species are replayed (recapitulated) in its embryology.

2. Some of my neuroscientist colleagues do not even recognize the concept of a neuromodulator; they view all of these chemicals as neurotransmitters that can be released into the synaptic cleft or into wider expanses of the brain.

7. Evolving Artificial Predictive Networks

1. This is a reference to the often-cited phrase of uncertain origins, *If all you have is a hammer, everything looks like a nail.*

2. In the article, they actually maximize the negative of FEP, also known as the *evidence lower bound* (ELBO).

3. Much of D'Arcy was inspired by the background reading for this book, and although fully implemented, the system is still in the early phases of testing.

8. Conclusion

1. Prior to the 2010s, this was an accusation that I often received, on behalf of AI, from engineers in other departments. Now, those same departments are creating their own AI professorships.

References

Abbott, Larry, Kurt Thoroughman, Astrid Prinz, Vatsala Thirumalai, and Eve Marder. 2003. "Activity-Dependent Modification of Intrinsic and Synaptic Conductances in Neurons and Rhythmic Networks." In *Modeling Neural Development,* edited by Arjen van Ooyen, 151–166. Cambridge, MA: MIT Press.

Ackley, David, Geoffrey Hinton, and Terrence Sejnowski. 1985. "A Learning Algorithm for Boltzmann Machines." *Cognitive Science* 9:147–169.

Ahmadi, Ahmadreza, and Jun Tani. 2019. "A Novel Predictive-Coding-Inspired Variational RNN Model for Online Prediction and Recognition." *Neural Computation* 31:2025–2074.

Albus, James. 1971. "A Theory of Cerebellar Function." *Mathematical Biosciences* 10:25–61.

Allman, John. 1999. *Evolving Brains.* New York: W.H. Freeman / Company.

Amit, Daniel. 2003. "Cortical Hebbian Modules." In *The Handbook of Brain Theory and Neural Networks,* edited by Michael Arbib, 285–290. Cambridge, MA: MIT Press.

Andersen, Per, Richard Morris, David Amaral, Tim Bliss, and John O'Keefe. 2007. *The Hippocampus Book.* New York: Oxford University Press.

Artola, Alain, Susanne Brocher, and Wolf Singer. 1990. "Different Voltage-Dependent Thresholds for Inducing Long-Term Depression and Long-Term Potentiation in Slices of Rat Visual Cortex." *Nature* 347:69–72.

Attneave, Fred. 1954. "Some Informational Aspects of Visual Perception." *Psychological Review* 61 (3): 183–193.

Baldominos, Alejandro, Yago Saez, and Pedro Isasi. 2020. "On the Automated, Evolutionary Design of Neural Networks: Past, Present and Future." *Neural Computing and Applications* 32:519–545.

Ballard, Dana. 2015. *Brain Computation as Hierarchical Abstraction.* Cambridge, MA: MIT Press.

Banzhaf, Wolfgang, Peter Nordin, Robert E. Keller, and Frank D. Francone. 1998. *Genetic Programming—An Introduction; On the Automatic Evolution of Computer Programs and Its Applications.* Burlington, MA: Morgan Kaufmann.

Barto, Andrew. 1995. "Adaptive Critics and the Basal Ganglia." In *Models of Information Processing in the Basal Ganglia,* edited by James Houk, Joel Davis, and David Beiser, 215–232. Cambridge, MA: MIT Press.

Bastos, Andre, Martin Usrey, Rick Adams, George Mangun, Pascal Fries, and Karl Friston. 2012. "Canonical Microcircuits for Predictive Coding." *Neuron* 76:695–711.

Bear, Mark, Barry Conners, and Michael Paradiso. 2001. *Neuroscience: Exploring the Brain.* 2nd ed. Baltimore, MD: Lippincott Williams / Wilkins.

Beer, Randall. 2003. "The Dynamics of Active Categorical Perception in an Evolved Model Agent." *Adaptive Behavior* 11 (4): 209–243.

Beer, Randall, and John Gallagher. 1982. "Evolving Dynamical Neural Networks for Adaptive Behavior." *Adaptive Behavior* 1 (1): 91–122.

Beer, Randall, and Paul Williams. 2015. "Information Processing and Dynamics in Minimally Cognitive Agents." *Cognitive Science* 39 (2): 1–38.

Bell, Curtis, David Bodznick, John Montgomery, and J. Bastian. 1997. "The Generation and Subtraction of Sensory Expectations within Cerebellum-Like Structures." *Brain, Behavior and Evolution* 50:17–31.

Bellmund, Jacob, Peter Gardenfors, Edvard Moser, and Christian Doeller. 2018. "Navigating Cognition: Spatial Codes for Human Thinking." *Science* 362 (July): eaat7666.

Bergquist, Harry, and Bengt Kallen. 1953. "On the Development of Neuromeres to Migration Areas in the Vertebrate Cerebral Tube." *Acta Anat (Basel)* 18:65–63.

Bogacz, Rafal. 2017. "A Tutorial on the Free-Energy Framework for Modelling Perception and Learning." *Journal of Mathematical Psychology* 76:198–211.

Bongard, Josh, and Rolf Pfeifer. 2001. "Repeated Structure and Dissociation of Genotypic and Phenotypic Complexity in Artificial Ontogeny." *Proceedings of the Genetic and Evolutionary Computation Conference (GECCO-2001)* 33:829–836.

Brockman, Greg, Vicki Cheung, Ludweig Pettersson, Jonas Schneider, John Schulman, Jie Tang, and Wojciech Zaremba. 2016. *Open AI Gym.* arXiv preprint:1606.01540.

Brooks, Rodney. 1999. *Cambrian Intelligence: The Early History of the New AI.* Cambridge, MA: MIT Press.

Burgess, Neil, and John O'Keefe. 2003. "Hippocampus: Spatial Models." In *The Handbook of Brain Theory and Neural Networks,* edited by Michael Arbib, 539–543. Cambridge, MA: MIT Press.

Buzsaki, Gyorgy. 2006. *Rhythms of the Brain.* New York: Oxford University Press.

Buzsaki, Gyorgy. 2019. *The Brain from Inside Out.* New York: Oxford University Press.

Carpenter, Gail, and Stephen Grossberg. 2003. "Adaptive Resonance Theory." In *The Handbook of Brain Theory and Neural Networks,* edited by Michael Arbib, 87–90. Cambridge, MA: MIT Press.

Clark, Andy. 2003. *Natural-Born Cyborgs.* New York: Oxford University Press.

Clark, Andy. 2016. *Surfing Uncertainty: Prediction, Action and the Embodied Mind.* Oxford: Oxford University Press.

Dayan, Peter, Geoffrey Hinton, Brendan Frey, and Radford Neal. 1995. "The Helmholtz Machine." *Neural Computation* 7 (5): 889–904.

Deacon, Terrence. 1998. *The Symbolic Species: The Co-evolution of Language and the Brain.* New York: W.W. Norton.

Dehaene, Stanislas. 1997. *The Number Sense.* New York: Oxford University Press.

Dickens, Charles. 1861. *Great Expectations.* London: Chapman / Hall.

Downing, Keith. 2007a. "Neuroscientific Implications for Situated and Embodied Artificial Intelligence." *Connection Science* 19 (1): 75–104.

Downing, Keith. 2007b. "Supplementing Evolutionary Developmental Systems with Abstract Models of Neurogenesis." In *GECCO '07: Proceedings of the 2007 Conference on Genetic and Evolutionary Computation,* 990–996. London.

Downing, Keith. 2009. "Predictive Models in the Brain." *Connection Science* 21 (1): 39–74.

Downing, Keith. 2015. *Intelligence Emerging: Adaptivity and Search in Evolving Neural Systems.* Cambridge, MA: MIT Press.

Doya, Kenji. 1999. "What Are the Computations of the Cerebellum, the Basal Ganglia, and the Cerebral Cortex?" *Neural Networks* 12:961–974.

Dunn, Felice, Martin Lankheet, and Fred Rieke. 2007. "Light Adaptation in Cone Vision Involves Switching between Receptor and Post-Receptor Sites." *Nature* 449:603–606.

Dusenbery, David. 1992. *Sensory Ecology: How Organisms Acquire and Respond to Information.* New York: W.H. Freeman.

Edelman, Gerald. 1987. *Neural Darwinism: The Theory of Neuronal Group Selection.* New York: Basic Books.

Edelman, Gerald. 1992. *Bright Air, Brilliant Fire: On the Matter of the Mind.* New York: Basic Books.

Edelman, Gerald, and Giulio Tononi. 2000. *A Universe of Consciousness.* New York: Basic Books.

Fahlman, Scott, and Christian Lebiere. 1997. "The Cascade-Correlation Learning Architecture." *Advances in Neural Information Processing Systems* 2 (October).

Finlay, Barbara, and Richard Darlington. 1995. "Linked Regularities in the Development and Evolution of Mammalian Brains." *Science* 268:1578–1584.

Fregnac, Yves. 2003. "Hebbian Synaptic Plasticity." In *The Handbook of Brain Theory and Neural Networks,* edited by Michael Arbib, 515–522. Cambridge, MA: MIT Press.

Friston, Karl. 2005. "A Theory of Cortical Responses." *Philosophical Transactions of the Royal Society B* 360: 815–836.

Friston, Karl. 2010. "A Free Energy Principle: A Unified Brain Theory?" *Nature Reviews Neuroscience* 11: 127–138.

Friston, Karl, James Kilner, and Lee Harrison. 2006. "A Free Energy Principle for the Brain." *Journal of Physiology—Paris* 100:70–87.

Fujita, M. 1982. "Adaptive Filter Model of the Cerebellum." *Biological Cybernetics* 45:195–206.

Fukushima, Kunihiko, and Sei Miyake. 1982. "Neocognitron: A New Algorithm for Pattern Recognition Tolerant of Deformations and Shifts in Position." *Pattern Recognition* 15:455–469.

Fuster, Joaquin. 2003. *Cortex and Mind: Unifying Cognition.* Oxford: Oxford University Press.

Gärdenfors, Peter. 2000. *Conceptual Spaces. The Geometry of Thought.* Cambridge, MA: MIT Press.

Glorot, Xavier, Antoine Bordes, and Yoshua Bengio. 2011. "Deep Sparse Rectifier Neural Networks." In *Proceedings 14th International Conference on Artificial Intelligence and Statistics,* 315–323. https://proceedings.mlr.press/v15/glorot11a/glorot11a.pdf.

Goodfellow, Ian, Yoshua Bengio, and Aaron Courville. 2016. *Deep Learning.* Cambridge, MA: MIT Press.

Gould, Stephen Jay. 1970. *Ontogeny and Phylogeny.* Cambridge, MA: Belknap Press of Harvard University Press.

Graybiel, Ann M., and Esen Saka. 2004. "The Basal Ganglia and the Control of Action." In *The Cognitive Neurosciences III,* edited by Michael S. Gazzaniga, 495–510. Cambridge, MA: MIT Press.

Grossberg, Stephen. 1969. "On Learning and Energy-Entropy Dependence in Recurrent and Nonrecurrent Signed Networks." *Journal of Statistical Physics* 1 (July): 319–350.

Hafting, Torkel, Marianne Fyhn, Sturla Molden, May-Britt Moser, and Edvard Moser. 2005. "Microstructure of a Spatial Map in the Entorhinal Cortex." *Nature* 436:801–806.

Harding, Simon, Julian Miller, and Wolfgang Banzhaf. 2009. "Evolution, Development and Learning Using Self-Modifying Cartesian Genetic Programming." In *Proceedings of the 11th Genetic and Evolutionary Computation Conference,* edited by Dirk Thierens et al., 699–706. Montreal, ACM.

Hasselmo, Michael, Bradley Wyble, and Erik Fransen. 2003. "Neuromodulation in Mammalian Nervous Systems." In *The Handbook of Brain Theory and Neural Networks,* edited by Michael Arbib, 761–765. Cambridge, MA: MIT Press.

Hawkins, Jeff. 2004. *On Intelligence.* New York: Henry Holt.

Hawkins, Jeff. 2021. *A Thousand Brains: A New Theory of Intelligence.* New York: Basic Books.

Hawkins, Jeff, and Subutal Ahmad. 2016. "Why Neurons Have Thousands of Synapses, a Theory of Sequence Memory in Neocortex." *Frontiers in Neural Circuits* 10 (March): 1–18.

Hawkins, Jeff, Subutal Ahmad, and Yuwei Cui. 2017. "A Theory of How Columns in the Neocortex Enable Learning the Structure of the World." *Frontiers in Neural Circuits* 11 (October): 1–13.

Hayes, Patrick. 1979. "The Naive Physics Manifesto." In *Expert Systems in the Micro-Electronic Age,* edited by Donald Michie, 761–765. Edinburgh: Edinburgh University Press.

Hebb, Donald. 1949. *The Organization of Behavior.* New York: John Wiley / Sons.

Hiesinger, Peter. 2021. *The Self-Assembling Brain: How Neural Networks Grow Smarter.* Princeton, NJ: Princeton University Press.

Hinton, Geoffrey. 2002. "Training Products of Experts by Minimizing Contrastive Divergence." *Neural Computation* 15 (8): 1771–1800.

Hinton, Geoffrey, Peter Dayan, Brendan Frey, and Radford Neal. 1995. "The Wake-Sleep Algorithm for Unsupervised Neural Networks." *Science* 268 (5214): 1158–1161.

Hinton, Geoffrey, and R. Salakhutdinov. 2006. "Reducing the Dimensionality of Data with Neural Networks." *Science* 313 (5214): 504–507.

Hochreiter, Sepp, and Jurgen Schmidhuber. 1997. "Long Short-Term Memory." *Neural Computation* 9 (8): 1735–1780.

Holland, John H. 1992. *Adaptation in Natural and Artificial Systems.* 2nd ed. Cambridge, MA: MIT Press.

Hopfield, John. 1982. "Neural Networks and Physical Systems with Emergent Collective Computational Abilities." *Proceedings of the National Academy of Sciences* 79:2554–2558.

Hopfield, John. 1984. "Neurons with Graded Response Properties Have Collective Computational Properties Like Those of Two-State Neurons." *Proceedings of the National Academy of Sciences* 81:3088–3092.

Hosoya, Toshihiko, Stephen Baccus, and Markus Meister. 2005. "Dynamic Predictive Coding by the Retina." *Nature* 436:71–77.

Houk, James, James Adams, and Andrew Barto. 1995. "A Model of How the Basal Ganglia Generate and Use Neural Signals That Predict Reinforcement." In *Models of Information Processing in the Basal Ganglia,* edited by James Houk, Joel Davis, and David Beiser, 249–270. Cambridge, MA: MIT Press.

Houk, James, Joel Davis, and David Beiser. 1995. *Models of Information Processing in the Basal Ganglia.* Cambridge, MA: MIT Press.

Izquierdo, Eduardo, and Randall Beer. 2013. "Connecting a Connectome to Behavior: An Ensemble of Neuroanatomical Models of *C. elegans* Klinotaxis." *PLoS Computational Biology* 9 (2): e1002890.

Kabadayi, Can, and Mathias Osvath. 2017. "Ravens Parallel Great Apes in Flexible Planning for Tool-Use and Bartering." *Science* 357 (July): 202–204.

Kandel, Eric, James Schwartz, and Thomas Jessell. 2000. *Principles of Neural Science.* New York: McGraw-Hill.

Kirschner, Marc W., and John C. Gerhart. 2005. *The Plausibility of Life: Resolving Darwin's Dilemma.* New Haven, CT: Yale University Press.

Kok, Peter, and Floris de Lange. 2015. "Predictive Coding in Sensory Cortex." In *An Introduction to Model-Based Cognitive Neuroscience,* edited by Birte Forstmann and Eric-Jan Wagenmakers, 221–244. New York: Springer.

Koza, John R. 1992. *Genetic Programming: On the Programming of Computers by Natural Selection.* Cambridge, MA: MIT Press.

Koza, John R., David Andre, Forrest H. Bennett III, and Martin Keane. 1999. *Genetic Programming 3: Darwinian Invention and Problem Solving.* Burlington, MA: Morgan Kaufman.

Krizhevsky, Alex. 2009. "Learning Multiple Layers of Features from Tiny Images." *Google Labs Tech Report.* www.cs.toronto.edu/~kriz/learning-features-2009-TR.pdf.

Kropff, Emilio, James Carmichael, May-Britt Moser, and Edvard Moser. 2015. "Speed Cells in the Medial Entorhinal Cortex." *Nature* 523 (July): 419–442.

Kuhn, Thomas. 1970. *The Structure of Scientific Revolutions.* Chicago, IL: University of Chicago Press.

Kurzweil, Ray. 2005. *The Singularity Is Near.* New York: Viking Press.

Kurzweil, Ray. 2012. *How to Create a Mind: The Secret of Human Thought Revealed.* New York: Viking Press.

Lake, Brenden, Ruslan Salakhutdinov, and Joshua Tenenbaum. 2015. "Human-Level Concept Learning through Probabilistic Program Induction." *Science* 350 (6266): 1332–1338.

Lakoff, George, and Mark Johnson. 1980. *Metaphors We Live By.* Chicago, IL: University of Chicago Press.

Lakoff, George, and Rafael Nunez. 2000. *Where Mathematics Comes From.* New York: Basic Books.

Langton, Christopher. 1989. "Artificial Life." In *Artificial Life: Proceedings of an Interdisciplinary Workshop on the Synthesis and Simulation of Living Systems,* edited by C. Langton, 1–49. Reading, MA: Addison-Wesley.

Larsch, Johannes, Steven Flavell, Qiang Liu, Andrew Gordus, Dirk Albrecht, and Cornelia Bargmann. 2015. "A Circuit for Gradient Climbing in C. elegans Chemotaxis." *Cell Reports* 12 (11): 1748–1760.

Larson, Erik. 2021. *The Myth of Artificial Intelligence: Why Computers Can't Think the Way We Do.* Cambridge, MA: Belknap Press of Harvard University Press.

LeCun, Yann, Yoshua Bengio, and Geoffrey Hinton. 2015. "Deep Learning." *Nature* 521:436–444.

LeCun, Yann, Bernhard Boser, John Denker, Donnie Henderson, R. Howard, Wayne Hubbard, and Lawrence Jackel. 1990. "Handwritten Digit Recognition with a Back-Propagation Network." In *Advances in Neural Information Processing Systems,* 396–404. Burlington, MA: Morgan Kaufmann.

LeDoux, Joseph. 2002. *Synaptic Self: How Our Brains Become Who We Are.* Middlesex, UK: Penguin Books.

Levine, Steve. 2017. "Artificial Intelligence Pioneer Says We Need to Start Over." *Axios* (September).

Levit, Mikhail, and Jeffry Stock. 2002. "Receptor Methylation Controls the Magnitude of Stimulus-Response Coupling in Bacterial Chemotaxis." *Journal of Biological Chemistry* 277 (29): 36760–36765.

Lillicrap, Timothy, Adam Santoro, Luke Marris, Colin Akermann, and Geoffrey Hinton. 2020. "Backpropagation and the Brain." *Nature Reviews Neuroscience* 21 (June): 335–346.

Lisman, John, and Anthony Grace. 2005. "The Hippocampal-VTA Loop: Controlling the Entry of Information into Long-Term Memory." *Neuron* 46:703–713.

Llinas, Rudolfo. 2001. *i of the Vortex.* Cambridge, MA: MIT Press.

Mackay, David. 2003. *Information Theory, Inference, and Learning Algorithms.* New York: Cambridge University Press.

Mackie, George. 1970. "Neuroid Conduction and the Evolution of Conducting Tissues." *Quarterly Review of Biology* 45:319–332.

Marcus, Mitchell, Mary Ann Marcinkiewicz, and Beatrice Santorini. 1993. "Building a Large Annotated Corpus of English: The Penn Treebank." *Computational Linguistics* 19 (2): 313–330.

Marr, David. 1969. "A Theory of Cerebellar Cortex." *Journal of Physiology* 202:437–470.

Marr, David. 1982. *Vision: A Computational Investigation into the Human Representation and Processing of Visual Information.* New York: Henry Holt.

Matsumoto, Takazumi, and Jun Tani. 2020. "Goal-Directed Planning for Habituated Agents by Active Inference Using a Variational Recurrent Neural Network." *Entropy* 22 (564): e22050564.

McNaughton, Bruce, Francesco Battaglia, Ole Jensen, Edvard Moser, and May-Britt Moser. 2006. "Path Integration and the Neural Basis of the 'Cognitive Map.'" *Nature Reviews Neuroscience* 7 (8): 663–678.

Mehta, Mayank. 2001. "Neuronal Dynamics of Predictive Coding." *Neuroscientist* 7 (6): 490–495.

Miikkulainen, Risto, Jason Zhi Liang, Elliot Meyerson, Aditya Rawal, Daniel Fink, Olivier Francon, Bala Raju, et al. 2017. "Evolving Deep Neural Networks." *CoRR* abs/1703.00548. arXiv: 1703.00548.

Miller, Julian. 2021. "DEMANNED: Designing Multiple ANNs via Evolved Developmental Neurons." *Artificial Life* 1 (1): 1–41.

Miller, Julian, and Wolfgang Banzhaf. 2003. "Evolving the Program for a Cell: From French Flags to Boolean Circuits." In *On Growth, Form and Computers,* edited by Sanjay Kumar and Peter Bentley, 278–301. Amsterdam: Elsevier Press.

Mitchell, Melanie. 2019. *Artificial Intelligence: A Guide for Thinking Humans.* London: Penguin Books.

Mitchell, Melanie, and Richard Belew. 1996. "Preface to Chapter 25." In *Adaptive Individuals in Evolving Populations: Models and Algorithms,* edited by Richard Belew and Melanie Mitchell, 443–445. Reading, MA: Addison-Wesley.

Mnih, Volodymry, Koray Kavukcuoglu, David Silver, Andrei Rosu, Joel Veness, Marc Bellemare, Alex Graves, Martin Riedmiller, Andreas Fidjeland, and Georg Ostrovski. 2015. "Human-Level Control through Deep Reinforcement Learning." *Nature* 518 (7540): 529–533.

Moriarty, David, and Risto Miikkulainen. 1997. "Forming Neural Networks through Efficient and Adaptive Coevolution." *Evolutionary Computation* 5 (4): 373–399.

Moser, Edvard, Yasser Roudi, Menno Witter, Clifford Kentros, Tobias Bonhoeffer, and May-Britt Moser. 2014. "Grid Cells and Cortical Representation." *Nature Reviews Neuroscience* 15 (July): 466–481.

Mountcastle, Vernon. 1998. *Perceptual Neuroscience: The Cerebral Cortex.* Cambridge, MA: Harvard University Press.

Mumford, David. 1992. "On the Computational Architecture of the Neocortex." *Biological Cybernets* 66:241–251.

Nakajima, K., Marc Maier, Peter Kirkwood, and Roger Lemon. 2000. "Striking Differences in Transmission of Coricospinal Excitation to Upper Limb Motoneurons in Two Primate Species." *Journal of Neurophysiology* 84:698–709.

Newell, Allen, and Herbert Simon. 1972. *Human Problem Solving.* Englewood Cliffs, NJ: Prentice Hall.

Nilsson, Nils. 1980. *Principles of Artificial Intelligence.* Palo Alto, CA: Tioga Publishers.

Nolfi, Stefano, and Dario Floreano. 2000. *Evolutionary Robotics: The Biology, Intelligence, and Technology of Self-Organizing Machines.* Cambridge, MA: MIT Press.

O'Keefe, John, and Jonathan Dostrovsky. 1971. "The Hippocampus as a Spatial Map: Preliminary Evidence from Unit Activity in the Freely-Moving Rat." *Journal of Brain Research* 34:171–175.

Oliver, Bernard. 1952. "Efficient Coding." *Bell Systems Technical Journal* 31:724–750.

Ooyen, Arjen van, Jaap van Pelt, Michael Corner, and Stanley Kater. 2003. "Activity-Dependent Neurite Outgrowth: Implications for Network Development and Neuronal Morphology." In *Modeling Neural Development,* edited by Arjen van Ooyen, 111–132. Cambridge, MA: MIT Press.

O'Reilly, Randall, and Yuko Munakata. 2000. *Computational Explorations in Cognitive Neuroscience.* Cambridge, MA: MIT Press.

Ouden, Hanneke den, Peter Kok, and Floris de Lange. 2012. "How Prediction Errors Shape Perception, Attention, and Motivation." *Frontiers in Psychology* 3 (548): 1–12.

Patton, Lydia. 2018. "Hermann von Helmholtz." In *The Stanford Encyclopedia of Philosophy,* Winter 2018, edited by Edward N. Zalta. Stanford, CA: Metaphysics Research Lab, Stanford University.

Petanjek, Zdravko, Milos Judas, Goran Simic, Mladen Rasin, Harry Uylings, Pasko Rakic, and Ivica Kostovic. 2011. "Extraordinary Neoteny of Synaptic Spines in the Human Prefrontal Cortex." *Proceedings of the National Academy of Sciences (PNAS)* 108 (32): 13181–13286.

Phattanasri, Phattanard, Hillel Chiel, and Randall Beer. 2007. "The Dynamics of Associative Learning in Evolved Model Circuits." *Adaptive Behavior* 15 (4): 377–396.

Porrill, John, and Paul Dean. 2016. "Cerebellar Adaptation and Supervised Learning in Motor Control." In *From Neuron to Cognition via Computational Neuroscience,* edited by Michael Arbib and James Bonaiuto, 617–647. Cambridge, MA: MIT Press.

Prescott, Tony, Kevin Gurney, and Peter Redgrave. 2003. "Basal Ganglia." In *The Handbook of Brain Theory and Neural Networks,* edited by Michael Arbib, 147–151. Cambridge, MA: MIT Press.

Puelles, Luis, and John Rubenstein. 1993. "Expression Patterns of Homeobox and Other Putative Regulatory Genes in the Embryonic Mouse Forebrain Suggest a Neuromeric Organization." *Trends in Neuroscience* 16: 472–479.

Quartz, Steven, and Terrence Sejnowski. 1997. "The Neural Basis of Cognitive Development: A Constructivist Manifesto." *Behavioral and Brain Sciences* (20): 537–596.

Queisser, Jeffrey, Minju Jung, Takazumi Matsumoto, and Jun Tani. 2021. "Emergence of Content-Agnostic Information Processing by a Robot Using Active Inference, Visual Attention, Working Memory and Planning." *Neural Computation* 33:2353–2407.

Rakic, Pasko. 2008. "Confusing Cortical Columns." *Proceedings of the National Academy of Sciences (PNAS)* 105 (34): 12099–12100.

Rao, Rajesh, and Dana Ballard. 1999. "Predictive Coding in the Visual Cortex: A Functional Interpretation of Some Extra-Classical Receptive-Field Effects." *Nature* 2 (1): 79–87.

Raut, Ryan, Abraham Snyder, and Marcus Raichle. 2020. "Hierarchical Dynamics as a Macroscopic Organizing Principle of the Human Brain." *Proceedings of the National Academy of Sciences (PNAS)* 117:20890–20897.

Real, Esteban, Chen Liang, David So, and Quoc Le. 2020. "AutoML-Zero: Evolving Machine Learning Algorithms from Scratch." *Proceedings of the 37th International Conference on Machine Learning (ICML)*. Vienna, Austria: PMLR.

Rodriguez, A., James Whitson, and Richard Granger. 2004. "Derivation and Analysis of Basic Computational Operations of Thalamocortical Circuits." *Journal of Cognitive Neuroscience* 16 (5): 856–877.

Rolls, Edmund, and Alessandro Treves. 1998. *Neural Networks and Brain Function.* New York: Oxford University Press.

Rumelhart, David, Geoffrey Hinton, and Ronald Williams. 1986. "Learning Internal Representations by Error Propagation." In *Parallel Distributed Processing: Explorations in the Microstructure of Cognition,* edited by David Rumelhart and James McClelland, 318–362. Cambridge, MA: MIT Press.

Salimans, Tim, Jonathan Ho, Xi Chen, Szymon Sidor, and Ilya Sutskever. 2017. "Evolution Strategies as a Scalable Alternative to Reinforcement Learning." *Open AI Lab Tech Report* (March). https://arxiv.org/pdf/1703.03864.pdf.

Sanes, Dan, Thomas Reh, and William Harris. 2006. *Development of the Nervous System.* Burlington, MA: Elsevier Academic Press.

Schneider, Gerald. 2014. *Brain Structure and Its Origins.* Cambridge, MA: MIT Press.

Schultz, Wolfram, Paul Apicella, Eugenio Scarnati, and Tomas Ljungberg. 1992. "Neural Activity in Monkey Ventral Striatum Related to the Expectation of Reward." *Journal of Neuroscience* 12 (12): 4595–4610.

Shatner, William. 2013. *Star Trek Convention—Saturday Night Live.* https://www.youtube.com/watch?v=Rqb4V9 GxaBo.

Shipp, Steward, Rick Adams, and Karl Friston. 2013. "Reflections on Agranular Architecture: Predictive Coding in the Motor Cortex." *Trends in Neuroscience:* 1–11. doi:10.1016/j.tins.2013.09.004.

Silver, David, Aja Huang, Chris Maddison, Arthur Guez, Laurent Sifre, George van den Driessche, Julian Schrittwieser, et al. 2016. "Mastering the Game of Go with Deep Neural Networks and Tree Search." *Nature* 529 (January): 484–503.

Silver, David, Thomas Hubert, Julian Schrittwieser, Ioannis Antonoglou, Matthew Lai, Arthur Guez, Marc Lanctot, et al. 2018. "A General Reinforcement Learning Algorithm That Masters Chess, Shogi and Go through Self-Play." *Science* 362 (6419): 1140–1144.

Silver, David, Julian Schrittwieser, Karen Simonyan, Ioannis Antonoglou, Aja Huang, Arthur Guez, Thomas Hubert, et al. 2017. "Mastering the Game of Go without Human Knowledge." *Nature* 550 (October): 354–371.

Sims, Karl. 1994. "Evolving 3D Morphology and Behavior by Competition." In *Artificial Life IV,* edited by R. Brooks and P. Maes, 28–39. Cambridge, MA: MIT Press.

Slocom, Andrew, Douglas Downey, and Randall Beer. 2000. "Further Experiments in the Evolution of Minimally-Cognitive Behavior: From Perceiving Affordances to Selective Attention." In *From Animals to Animats 6:*

Proceedings of the Sixth International Conference on Simulation of Adaptive Behavior, edited by J. Meyer, A. Berthoz, D. Floreano, H. Roitblat, and S. Wilson, 430–439. Cambridge, MA: MIT Press.

Smolensky, Paul. 1986. "Information Processing in Dynamical Systems: Foundations of Harmony Theory." In *Parallel Distributed Processing: Explorations in the Microstructure of Cognition,* edited by David Rumelhart and James McClelland, 194–263. Cambridge, MA: MIT Press.

Soltoggio, Andrea, John Bullinaria, Claudio Mattiussi, Peter Durr, and Dario Floreano. 2008. "Evolutionary Advantages of Neuromodulated Plasticity in Dynamic, Reward-based Scenarios." In *Artificial Life XI,* edited by S. Bollock, J. Noble, R. Watson, and M. Bedau, 569–576. Cambridge, MA: MIT Press.

Soltoggio, Andrea, Kenneth Stanley, and Sebastian Risi. 2018. "Born to Learn: The Inspiration, Progress, and Future of Evolved Plastic Artificial Neural Networks." *Neural Networks* 1:48–67.

Song, Sen, Kenneth Miller, and Larry Abbott. 2000. "Competitive Hebbian Learning through Spike-Timing-Dependent Synaptic Plasticity." *Nature Neuroscience* 3 (9): 919–926.

Song, Yuhang, Thomas Lukasiewicz, Zhenghua Xu, and Rafal Bogacz. 2020. "Can the Brain Do Backpropagation? Exact Implementation of Backpropagation in Predictive Coding Networks." *Advances in Neural Information Processing Systems* 33:22566–22579.

Spratling, Michael. 2008. "Reconciling Predictive Coding and Biased Competition Models of Cortical Function." *Frontiers in Computational Neuroscience* 2 (4): 1–8.

Spratling, Michael. 2017. "A Review of Predictive Coding Algorithms." *Brain and Cognition* 112 (March): 92–97.

Spratling, Michael. 2019. "Fitting Predictive Coding to the Neurophysiological Data." *Brain Research* (October): 1–12.

Squire, Larry, and Stuart Zola. 1996. "Structure and Function of Declarative and Nondeclarative Memory Systems." *Genetic Programming and Evolvable Machines* 93:13515–13522.

Srinivasan, Mandyam, Simon Laughlin, and A. Dubs. 1982. "Predictive Coding: A Fresh View of Inhibition in the Retina." *Proceedings of the Royal Society of London B* 216:427–459.

Stanley, Kenneth. 2007. "Compositional Pattern Producing Networks: A Novel Abstraction of Development." *Genetic Programming and Evolvable Machines: Special Issue on Developmental Systems* 8 (2): 131–162.

Stanley, Kenneth, Jeff Clune, Jeff Lehman, and Risto Miikkulainen. 2019. "Designing Neural Networks through Neuroevolution." *Nature Machine Intelligence* 1:24–35.

Stanley, Kenneth, and Risto Miikkulainen. 2002. "Evolving Neural Networks through Augmenting Topologies." *Evolutionary Computation* 10 (2): 99–127.

Stanley, Kenneth, and Risto Miikkulainen. 2003. "A Taxonomy for Artificial Embryogeny." *Artificial Life* 9 (2): 93–130.

Steels, Luc. 2003. "Intelligence with Representation." *Philosophical Transactions: Mathematical, Physical and Engineering Sciences* 361 (1811): 2381–2395.

Sterling, Peter, and Simon Laughlin. 2015. *Principles of Neural Design.* Cambridge, MA: MIT Press.

Stone, James. 2018. *Principles of Neural Information Theory.* Sheffield, UK: Sebtel Press.

Stone, James. 2020. *Artificial Intelligence Engines.* Sheffield, UK: Sebtel Press.

Striedter, Georg. 2005. *Principles of Brain Evolution.* Sunderland, MA: Sinauer Associates.

Sutton, Richard S., and Andrew G. Barto. 1998. *Reinforcement Learning: An Introduction.* Cambridge, MA: MIT Press.

Sutton, Richard S., and Andrew G. Barto. 2018. *Reinforcement Learning: An Introduction.* Cambridge, MA: MIT Press.

Tani, Jun. 2014. "Self-Organization and Compositionality in Cognitive Brains: A Neurorobotics Study." *Proceedings of the IEEE* 102 (4): 586–605.

Tani, Jun. 2017. *Exploring Robotic Minds.* New York: Oxford University Press.

Thomson, Alex, and Peter Bannister. 2003. "Interlaminar Connections in the Neocortex." *Cerebral Cortex* 13 (1): 5–14.

Thompson, D'Arcy. 1992. *On Growth and Form.* Mineola, NY: Cambridge University Press.

Tripp, Bryan, and Chris Eliasmith. 2010. "Population Models of Temporal Differentiation." *Neural Computation* 22 (3): 621–659.

Wallenstein, Gene, Howard Eichenbaum, and Michael Hasselmo. 1998. "The Hippocampus as an Associator of Discontiguous Events." *Trends in Neuroscience* 21 (8): 317–323.

Werbos, Paul. 1990. "Backpropagation through Time: What It Does and How to Do It." *Proceedings of the IEEE* 78 (10): 1550–1560.

Whittington, James, and Rafal Bogacz. 2017. "An Approximation of the Error Backpropagation Algorithm in a Predictive Coding Network with Local Hebbian Synaptic Plasticity." *Neural Computation* 29:1229–1262.

Whittington, James, and Rafal Bogacz. 2019. "Theories of Error Back-Propagation in the Brain." *Trends in Cognitive Sciences* 23 (3): 235–250.

Wierstra, Daan, Tom Schaul, Tobias Glasmachers, Yi Sun, Jan Peters, and Jurgen Schmidhuber. 2014. "Natural Evolution Strategies." *Journal of Machine Learning Research* 15:949–980.

Wolpert, Daniel, R. Miall, and Mitsuo Kawato. 1998. "Internal Models in the Cerebellum." *Trends in Cognitive Sciences* 2 (9): 338–347.

Yamashita, Yuichi, and Jun Tani. 1996. "Emergence of a Functional Hierarchy in a Multiple Timescale Neural Network Model: A Humanoid Robot Experiment." *PLoS Computational Biology* 4 (11): e37843.

Yao, Xin. 1999. "Evolving Artificial Neural Networks." *Proceedings of the IEEE* 87 (9): 1423–1447.

Yuille, Alan, and Daniel Kersten. 2006. "Vision as Bayesian Inference: Analysis by Synthesis?" *Trends in Cognitive Science* 10 (7): 301–308.

Index